Lecture Notes in Mathematics

Edited by A. Dold and B. Eckmann

1343

Hanno Ulrich



# Fixed Point Theory of Parametrized Equivariant Maps

Springer-Verlag
Berlin Heidelberg New York London Paris Tokyo

**Author**

Hanno Ulrich
IBM Research and Development
Schönaicher Straße 220, 7030 Böblingen, FRG

Mathematics Subject Classification (1980): 55-02; 19L47, 54C55, 54D50, 54F45, 55M20, 55N91, 55P91, 55R65, 55R91, 57R35

ISBN 3-540-50187-8 Springer-Verlag Berlin Heidelberg New York
ISBN 0-387-50187-8 Springer-Verlag New York Berlin Heidelberg



Printing and binding: Druckhaus Beltz, Hemsbach/Bergstr.
2146/3140-543210

*To my Parents*

# Preface

It is a pleasure for me to acknowledge my debt to all those who have contributed to this book.

My interest in fixed point theory originates from lectures given by Albrecht Dold at the university of Heidelberg. I am especially thankful to him and to Dieter Puppe who taught me most of what I know about algebraic topology and homotopy theory. Concerning the book's actual contents, I could further profit from various papers of Tammo tom Dieck and his pupils treating related topics.

I am very grateful to my friends Mónica Clapp and Carlos Prieto for their repeated encouragements to publish this text, and I am obliged to the Springer-Verlag for accepting it into its Lecture Notes of Mathematics series. The project, however, could not have been realized without the generous assistance of IBM. I was impressed how the IBM research and development laboratory in Böblingen promoted the scientific interests of one of its employees which, by no means, were related to the company's vital business interests.

IBM subsidized the translation of the German original done by Andreas Schroth and Christiane Vierthaler. I owe both of them sincere thanks for their careful efforts. The tiresome work of typing the manuscript and formatting the text to its final layout was considerably facilitated by IBM's text processing tools. For printing, an IBM 3820 all-points-addressable printer was at my disposal. I am very much indebted to my managers Karl Niebel and Günther Sonntag who supported the publication and who gave me the time to complete it.

Finally, I wish to thank my wife for her patient understanding and for her moral help over the past year.

Tübingen, fall 1987 *Hanno Ulrich*

# Contents

# Introduction

In this book, we are developing an *algebraic fixed point theory* for *equivariant maps,* that is for maps with symmetry properties.

So, let $E$ be a topological space with a continuous action of some topological group $G$, a *$G$-space* in short, and let $f$ be a partially defined $G$-transformation on $E$: I.e. the domain of $f$ has to be a $G$-subspace of $E$ and $f$ is to respect the action of $G$. Maps are always understood to be continuous.

More generally, we consider *continuous $G$-families* $f = \{f_b\colon V_b \to E_b,\ b \in B\}$ of such maps where $G$ may act non-trivially on the parameter space $B$: An element $g \in G$ transports the fibre $f_b$ of $f$ to the fibre over $gb \in B$ and for $v \in V_b$, $f_{gb}(gv)$ equals $gf_b(v)$. If we let $E$ denote the union of all $E_b$, $b \in B$, then $G$ acts on $E$ and $p\colon E \to B$, $p^{-1}(b) := E_b$, is a $G$-map. $V := \bigcup_{b\in B} V_b$ is a $G$-subspace of $E$ over $B$ and $f$ becomes a *vertical $G$-map*

$$\begin{array}{ccc} V & \xrightarrow{\ f\ } & E \\ & {\scriptstyle p}\searrow \quad \swarrow{\scriptstyle p} & \\ & B & \end{array}$$

with $pf(v) = p(v)$, $f(gv) = gf(v)$, and $p(gv) = gp(v)$. We call $f$ a *$G$-fixed point situation over $B$* if its fixed point set $\mathrm{Fix}(f) := \{v \in V, f(v) = v\}$ lies properly over $B$, if $V$ is open in $E$, and if $p$ is a $G$-$\mathrm{ENR}_B$, in words a *$G$-euclidian neighbourhood retract over $B$*. So, there exist a *$G$-module $M$*, i.e. a linear representation of $G$ in some euclidean space $M = \mathbb{R}^n$, and a vertical $G$-embedding of $p$ into the projection $B \times M \to B$ such that therein, $p$ is a vertical $G$-retract of some $G$-neighbourhood. We assume that the base $B$ *is paracompact and compactly generated*. Then the fixed point set of $f$ lies properly over $B$ if and only if it is closed and fibrewise uniformly bounded as a subspace of $B \times M$.

In [Dold 2], any non-equivariant fixed point situation $f$ over $B$ gets assigned a *fixed point index* $I(f)$ in the *zeroth stable cohomotopy group* $\pi_s^0(B)$ of $B$ constructed as follows: Each fibre of $f$ has a Hopf index $I(f_b) \in \mathbb{Z}$ represented by a transformation $i(f_b)$ of the pointed $n$-sphere, or of the pair $(\mathbb{R}^n, \mathbb{R}^n - 0)$, of degree $I(f_b)$. The family of all these transformations $i(f_b)$ can be constructed so as to depend continuously on the parameter $b \in B$. It then constitutes a map

$$i(f)\colon B \times (\mathbb{R}^n, \mathbb{R}^n - 0) \to (\mathbb{R}^n, \mathbb{R}^n - 0)$$

which represents the fixed point index $I(f) \in \pi_s^0(B)$. When $f$ is a $G$-fixed point situation with $G$ a compact (Lie) group, the analogous construction yields a $G$-map

$$i_G(f): B \times (M, M-0) \to (M, M-0).$$

which will represent the *$G$-fixed point index $I_G(f)$ in the zeroth stable $G$-cohomotopy group* $\pi_G^0(B)$. The fixed point index of $f$ in any other *multiplicative $G$-cohomology theory $h_G$* is the image of $I_G(f)$ under the degree map $u: \pi_G{}^*(B) \to h_G{}^*(B)$. The $h_G$-index of the identity on a proper $G$-ENR$_B$ $p$ will be called the *$h_G$-characteristic of $p$*, denoted by $\chi_{h_G}(p)$. Two $G$-fixed point situations $f_0$ and $f_1$ over $B$ are said to be *equivalent* if they can be connected through a $G$-fixed point situation $f = \{f_t\}$ over $B \times [0, 1]$. The equivalence classes form a ring with unit element, the *$G$-fixed point ring $Fix_G(B)$*. By virtue of its basic properties, the fixed point index in a multiplicative $G$-cohomology theory $h_G$ defines a homomorphism of unitary rings

$$I_{h_G}: Fix_G(B) \to h_G^0(B).$$

We show that $I_G := I_{\pi_G}$ is an *isomorphism.* It will be indicated only how stable $G$-cohomotopy classes of non-trivial degree $\alpha$ can be realized by $G$-fixed point situations of degree $\alpha$ where $\alpha$ is an element of the real representation ring of $G$.
By interpreting maps between equivariant cohomotopy groups in terms of equivariant fixed point theory, we attain a geometrical view of the *equivariant group completion map* and thus of the mapping in the *Segal conjecture.* And by means of the *fixed point transfer in equivariant K-theory,* we can describe what is meant by *induced representations in the category of compact Lie groups.*

The *sum formula* in Chapter III, Section 5 is a crucial tool for calculating the equivariant fixed point index over a point: The $h_G$-index of a $G$-fixed point situation $f$ decomposes into an integral linear combination of the $h_G$-characteristics of the orbit types of $G$ around its fixed point set, that is

$$I_{h_G}(f) = \sum_{(H)} n_H(f) \cdot \chi_{h_G}(G/H).$$

$\chi_{h_G}$ vanishes on all orbit types $(G/H)$ whose $G$-automorphism group $W(H) = N(H)/H$ is not finite whereas otherwise, $n_H(f)$ is the Hopf index

$$n_H(f) = \left(I(f^H) - I(f^{\underline{H}})\right) / |W(H)| \in \mathbb{Z}.$$

$f^H$ and $f^{\underline{H}}$ denote the fixed point situations induced by $f$ in the $H$-fixed point spaces $E^H := \{x \in E,\ hx = x \text{ for all } h \in H\}$ and $E^{\underline{H}} := \bigcup_{K>H} E^K$. For example, if $S \leq G$ is a torus, then the *Hopf index* $I(f)$ of $f$ equals $I(f^S)$, and if $G$ is a finite $p$-group, it is congruent to $I(f^G)$ modulo $p$. Therefore, the Hopf index of any odd transformation of a sphere is even. This is the *Borsuk-Ulam theorem*. Generalizations follow at once by means of the sum formula.
We will further discuss the relation between the index of a $G$-fixed point situation $f$ and that of the map $f/G$ induced on orbit spaces - provided of course, the latter's fixed point set is compact. To illustrate the ideas developed, we derive three results of A. Weil, H. Hopf, and D.H. Gottlieb which nowadays are folklore in the theory of compact Lie groups.

For the identity on a compact $G$-ENR $E$, the sum formula takes the form

$$\chi_{h_G}(E) = \sum_{(H)} \chi_c(E_{(H)}/G) \cdot \chi_{h_G}(G/H)$$

where $E_{(H)} \subset E$ is the $G$-subspace consisting of all points on orbits of type $(G/H)$ and $\chi_c$ denotes the Euler-Poincaré characteristic in singular or Čech cohomology with compact support. Thus the *Euler-Poincaré characteristic of* $E$ decomposes as

$$\chi(E) = \sum_{(H)} \chi_c(E_{(H)}).$$

Regarding again the sum formula for general $f$, the last result might suggest that the $(H)^{\text{th}}$ term would be counting the fixed points of $f$ in $E_{(H)}$. This, however, is not correct as a fixed point of $f$ living in $E^K$ with $K > H$ may be counted by $f^K$ with a multiplicity different from that counted by $f^H$.

Interpreted in stable $G$-cohomotopy theory, the sum formula purports that every $G$-fixed point situation over a point is equivalent to the identity on a suitable compact $G$-ENR. The *coefficient ring* $F(G) := Fix_G(\text{pt})$ of $G$-fixed point theory even coincides with the *Burnside ring* $A(G)$ of $G$ and our isomorphism $I_G: F(G) \to \pi^0_G(\text{pt})$ provides a *geometrical view of the tom Dieck isomorphism* $A(G) \cong \pi_G^0(\mathrm{pt})$ [tom Dieck 2]. For, $A(G)$ is defined as the set of equivalence classes of compact $G$-ENRs under the equivalence relation $E \sim E'$ if $\chi(E^H) = \chi(E'^H)$ for all $H \leq G$; and $f \mapsto (I(f^H))$ induces an injective ring homomorphism $F(G) \to \prod_{H \leq G} \mathbb{Z}$. Indeed, using the sum formula, we see that already the corresponding homomorphism

$$I^*: F(G) \to \mathbb{Z}^{\Phi(G)}, \quad [f] \mapsto (I(f^H))$$

is injective where $\Phi(G)$ denotes the set of those orbit types of $G$ whose $G$-automorphism group is finite. By inverting the matrix of $I^*$ - as an abelian group, $F(G)$ is free on $\Phi(G)$ - we can *describe the image of* $F(G)$ *in* $\mathbb{Z}^{\Phi(G)}$ *by relations and congruences:* For every finite subset $\mathcal{H} \subset \Phi(G)$, there exists a relation specifying in $\mathbb{Z}^{\Phi(G)}$ a free subgroup $C_{\mathcal{H}}$ of rank $|\mathcal{H}|$ which contains the image of the subgroup of $F(G)$ free on $\mathcal{H}$, and further a set of congruences determining this image in $C_{\mathcal{H}}$.

The inverse of $I^*$ is a three-dimensional matrix M over $\mathbb{Z}$. We present an explicit recipe to calculate its entries. For $G$ finite, M reduces to a two-dimensional matrix. And if $G$ is a finite cyclic group generated by some some $g$, M provides us with congruences between the Hopf indices of the iterates $g^k$ of $g$.

In the union of all $C_{\mathcal{H}}$, we can describe the image of $F(G)$ by a closed set of congruences - even if $G$ is not finite. It is derived from the following *congruence* holding for finite group actions:

$$\sum_{g \in G} I(f^g) \equiv 0 \mod |G|.$$

The superscript "$g$" stands for the subgroup of $G$ generated by $g$. This relation, in turn, will be deduced by several ways: First, it is an immediate consequence of the sum formula which, for the identity on a compact $G$-ENR, reads

$$\sum_{g \in G} \chi(E^g) = \chi(E/G)\,|G|.$$

Second, we find an elementary approach by investigating local Hopf indices, and finally, we will encounter it again in the last section when we are analyzing the fixed point index in equivariant $K$-theory.

For *in equivariant K-theory*, the index of a $G$-fixed point situation $f$ over a point is an element of the complex representation ring $R(G) = K_G(\mathrm{pt})$ of $G$. Employing the sum formula, we can calculate its character function as

$$I_{K_G}(f): G \to \mathbb{Z}, \quad g \mapsto I(f^g).$$

In particular, the $K_G$-characteristic of a compact $G$-ENR is the virtual representation

$$\chi_{K_G}(E) = \sum (-1)^i H^i(E; \mathbb{C})$$

of $G$ over $\mathbb{C}$. For a finite $G$-set $E$, this is the permutation representation $\mathbb{C}(E)$ of $G$.
We end our discourse with a description of the $K_G$-*transfer* in terms of *Atiyah's topological index homomorphism* $t$-ind and a final application of the *Atiyah-Singer index formula*.

These are the contents of **Chapter III.** For the sake of completeness, we have enclosed an **Appendix** on *proper maps*.

The results from the *theory of compact transformation groups* used in this book are listed in **Chapter I.** There we also prove a simplified version of the *equivariant transversality theorem* required lateron when we will discuss local Hopf indices.

$G$-ENR$_B$s are discussed in **Chapter II.** Throughout, $G$ will be a compact group equipped with a Lie-structure if necessary. In the results stated, $B$ is assumed to be paracompact.

We confirm the *properties of* $G$-ENR$_B$s well-known for the non-equivariant case. So, a closed vertical $G$-subspace $q$ of a $G$-ENR$_B$ $p$ is a $G$-ENR$_B$ if and only if the inclusion $q \to p$ is a *$G$-cofibration over $B$*. Proper $G$-ENR$_B$s are $G$-fibrations. Arbitrary $G$-ENR$_B$s, however, only enjoy the *$G$-beginning covering homotopy property* $G$-BCHP: In general, $G$-homotopies on the base can be $G$-lifted just a bit.
It turns out that a vertical $G$-space $p: E \to B$ over a $G$-ENR is a $G$-ENR$_B$ if and only if it has the $G$-BCHP and $E$ is a $G$-ENR. This allows us to confirm a conjecture proposed in [Dold 3] for the non-equivariant case: Over a $G$-ENR $B$, a $G$-ENR$_B$ is characterized as a vertical $G$-space which has the $G$-BCHP and all of whose fibres are $G$-ENRs.

A smooth manifold is an ENR and a submersion of smooth manifolds is a vertical ENR since it comes with a family of local cross sections and hence is a $C^0$-submersion. We formulate this equivariantly in three versions of different strength and investigate how such *$G$-$C^0$-submersions* are related to spaces having the $G$-BCHP.
While the $G$-BCHP distinguishes vertical $G$-ENRs over a $G$-ENR, a vertical $G$-space whose total space is a $G$-ENR, proves to be a $G$-ENR$_B$ if and if only it is a $G$-$C^0$-submersion. We will see however that in a $G$-ENR$_B$ $p$ with surjective mappings $p^H$ for all $H \leq G$, the base is already a $G$-ENR if - and hence only if - the total space is one.

To prove the last two results, we employ a *vertical generalization of the Jaworowski criterion for* $G$-ENRs with a compact Lie group: A closed $G$-subspace $p: E \to B$ of a vertically trivial $G$-ENR$_B$ $B \times M \to B$ over a sufficiently nice base is a $G$-ENR$_B$ if and only if $p^H$ is a vertical ENR over $B^H$ for each orbit type $(H)$ on $B \times M - E$.

Examples of $G$-ENR$_B$s are provided by *locally trivial fibre bundles with action of a compact Lie group $G$*. If such a fibre bundle is locally trivial in an equivariant sense, it proves to be an extra strong $G$-$C^0$-submersion and hence a $G$-fibration. This is the homotopy theorem in [tom Dieck 1].
If $p$ is a $G$-locally trivial bundle of finite type over a suitable base, we can specify ENR-conditions for the non-equivariant fibre $F$ which identify $p$ as a $G$-ENR$_B$. In $G$-vector bundles, for instance, these conditions are satisfied. From this, we can derive an ENR-condition for $F$ sufficient to make $p$ a $G$-ENR$_B$ even if $p$ is not $G$-locally trivial.
For illustration, we study the projection of a $G$-ENR $E$ onto its orbit space: It will be a vertical $G$-ENR if and only if there is only one orbit type on $E$ - locally at least.

We end the chapter with various examples. From an inductive criterion, we deduce that smooth $G$-manifolds of finite orbit structure are $G$-ENRs. Hence, a $G$-$C^\infty$-submersion of such manifolds is a vertical $G$-ENR, for it is a $G$-$C^0$-submersion. Finally, with regard to our sum formula, we show that the saturations $p^{(H)} := G \cdot p^H$ and $p^{\underline{(H)}} := G \cdot p^{\underline{H}}$ of the $H$-fixed point spaces in a $G$-ENR$_B$ $p$ are $G$-ENR$_B$s for themselves.

Further comments will be found in the extensive introductions preceding each chapter. We use the symbol □ to mark the end of a proof whereas the symbol ⊡ indicates that the proof remains to be completed in a subsequent section.

# CHAPTER I
# Preliminaries on Group Actions

The **first two sections** establish the notational conventions and list some classical results on compact transformation groups. For details, we refer the reader to the books [Bredon] and [Palais 2]. In addition, we provide the following two results required in the subsequent chapters.
Let $G$ be a *compact Lie group*. The set of conjugacy classes of closed subgroups of $G$ comes with a natural partial ordering and we first show how it may be refined to a well-ordering. Second, some useful *sum formulae of R.S. Palais*, relating the dimension of a $G$-space to that of its orbit space, will be generalized from separable, metrizable $G$-spaces to just metrizable and even to paracompact, perfectly normal ones.

The entire **Section 3** has been devoted to derive a simplified version of the *equivariant transversality theorem* following along the lines of [Hausschild 1].

In the final **Section 4**, we will sketch the local characterization of *G-cofibrations* and *G-fibrations* whereby $G$ is merely a compact group.

## 1. Some Set Theory on Compact Lie Groups

**1.1** Let $G$ always denote a *compact topological group*. Continuous vector valued functions on $G$ can thus be integrated. If necessary, $G$ comes with the structure of a Lie group.
A *subgroup* $H \leq G$ is always to be a *closed* subspace of $G$. Hence, $H$ as well as $G$ is a compact (Lie) group. Let $(H)$ denote the *conjugacy class* of $H$ in $G$. $(H) \leq (K)$ means that $H$ is subconjugate to $K$. $G/H$ stands for the set of *left cosets* $gH$ of $H$ in $G$. By $W(H)$, we denote the *Weyl group* $N(H)/H$ where $N(H)$ is the *normalizer* of $H$ in $G$. We write $H \trianglelefteq G$ in case $H$ is a *normal* subgroup of $G$. $\langle g \rangle \leq G$ is the topological subgroup *generated* by $g \in G$. Its order is denoted by $|g|$.

If $G$ is a compact Lie group, then, up to conjugation, the number of subgroups $H \leq G$ is countable ([Palais 2]). In this case, therefore, the partial ordering " $\leq$ " on the set of

conjugacy classes in $G$ can readily be refined to a total ordering - simply by a numeration with real numbers:
One enumerates the conjugacy classes $(H)$ with natural numbers and places $(H_1)$ at the origin, say. If $(H_2)$ and $(H_1)$ are incomparable, we assign to $(H_2)$ any real number different from zero. Otherwise, $(H_2)$ gets a positive real number in case $(H_2) > (H_1)$ and a negative one else. We then place $(H_3)$ at a suitable point and continue.
Of course, we will not get a well-ordering that way. But as indicated, we can show:

**1.2 Theorem.** *Let $G$ be a compact Lie group. The partial ordering "$\leq$" on the set of conjugacy classes $\{(H), H \leq G\}$ in $G$ can be refined to a well-ordering "$\lesssim$".*

PROOF. Let $\mathscr{H}$ be any non-empty set of conjugacy classes in $G$ and select some $(H_1) \in \mathscr{H}$. If $(H_1)$ is not minimal with respect to the partial ordering "$\leq$" on $\mathscr{H}$, then there exists a subgroup $H_2 < H_1$ with $(H_2) \in \mathscr{H}$. If $(H_2)$ neither is minimal in $\mathscr{H}$, we find some $(H_3)$ in $\mathscr{H}$ such that $H_3 < H_2 < H_1$. This chain will cease eventually since, as a proper submanifold, $H_{i+1}$ has less components or a smaller dimension than $H_i$. Hence, there is a minimal element in $\mathscr{H}$.
In other words, the set of conjugacy classes in $G$ satisfies the descending chain condition. The assertion now follows from the next lemma. □

**1.3 Lemma.** *Let $(X, \omega)$ be a non-empty, partially ordered set. The ordering $\omega$ can be refined to a well-ordering on $X$ if and only if $\omega$ satisfies the descending chain condition.*

PROOF. The descending chain condition on $\omega$ is necessary because a totally ordered set is well-ordered if and only if it satisfies the descending chain condition.
Conversely, we consider the set $W$ of all well-orderings $w_A$ refining $\omega$ which are defined on subsets $A \subset X$ with the property that $a \in A$ implies $x \in A$ for all $\omega$-predecessors of $a$.
By calling $w_A$ smaller than $w_B$ if $w_A$ is an initial segment in $w_B$, we equip $W$ with an inductive ordering: For, any chain $\{w_A\}$ in $W$ defines, on the union of all its domains $A$, a well-ordering belonging to $W$ which yields an upper bound for the chain. Clearly, $W$ is non-empty - the empty set for instance belongs to $W$. Hence, Zorn's Lemma provides a maximal element $w_M$ in $W$.
If $X - M$ were non-empty, then, because of the descending chain condition, we could find therein an element $x$ minimal with respect to $\omega$. Putting $x$ behind all of $M$, we would define a well-ordering in $W$ strictly geater than $w_M$: For, all $\omega$-predecessors of $x$ were in $M$ because of the minimum choice of $x$ while, by definition, none of the $\omega$-successors of $x$ could belong to $M$. □

# 2. On the Topology of Spaces with Group Action

**2.1** Throughout the section, let $G$ be a *compact group.*

By a $G$-space, we understand a topological space $X$ equipped with a continuous action of $G$ from the left, i.e. with a continuous multiplication $(-\cdot-): G\times X\to X$. An equivariant mapping of $X$ - a *G-map* in short - is a continuous map $f$ from $X$ to some other $G$-space respecting the group action: $f(gx)=gf(x)$. By a *G-transformation* of $X$, we mean an equivariant self mapping of $X$.
$G/H$ for example is a $G$-space for any $H\leq G$. The set of its $G$-transformations is the Weyl group $W(H)=N(H)/H$.

**2.2** For any $G$-subspace $A\subset X$ the closure $\overline{A}$, the interior $\mathring{A}$, and the complement $X-A$ are $G$-subspaces of $X$. If $A$ is the zero-set of some function $\tau: X\to[0, 1]$, we may assume that $\tau$ is a $G$-function. For otherwise, we integrate $\tau$ over $G$.
The *orbit* of a point $x\in X$ is denoted by $Gx$ and the *isotropy subgroup* of $G$ at $x$ by $G_x$. Since $G$ is compact, the natural map $G/G_x\to Gx$, $gG_x\mapsto gx$ is a $G$-homeomorphism.
Every neighbourhood of an orbit in $X$ contains a $G$-neighbourhood since $G$ is compact. If $X$ is a Hausdorff space, the closed $G$-neighbourhoods constitute neighbourhood bases around the orbits of $G$, for the action of a compact group is a closed map.

**2.3** The *orbit space* of $X$ will be denoted by $X/G$. The projection $X\to X/G$ is proper, hence in particular closed, and open. Therefore, $X/G$ is a Hausdorff, a (completely) regular, or a (perfectly) normal space, just as $X$ is. Furthermore, $X/G$ will be (locally) compact, Lindelöf-compact, or compactly generated if and only if $X$ is so. In particular, when $X$ is separable and metrizable, then so is $X/G$. According to a theorem of E. Michael, $X/G$ is paracompact exactly if $X$ is. All these results are detailed in the book [Engelking].

**2.4** By the *type of an orbit* we mean its equivariant topological type. The set of orbit types on $X$ comes with a partial ordering, namely type$(Gx)\leq$ type$(Gy)$ if there is a $G$-map from $Gx$ to $Gy$. This in turn holds if and only if the isotropy group $G_x$ is subconjugate to $G_y$, i.e. if $(G_x)\leq(G_y)$.
Without any harm, we may therefore identify the orbit type of $Gx$ with the conjugacy class of $G_x$.

**2.5** For $H \leq G$, let $X^H$ denote the $H$-fixed point set of $X$, i.e. the set of all points in $X$ whose isotropy group contains $H$. $X^H$ carries an obvious action of $N(H)$ or $W(H)$. We write $X^{(H)}$ for the $G$-saturation $GX^H$. It consists of all points in $X$ whose isotropy group $H$ is subconjugate to.
By $X_{(H)}$, we denote the set of all points on orbits of type $(G/H)$, and by $X_H \subset X_{(H)}$ the subspace of all points whose isotropy group is exactly $H$. The complement $X^{\underline{(H)}}$ of $X_{(H)}$ in $X^{(H)}$ consists of all those points whose isotropy group is of type strictly greater than $(H)$.

**2.6** A $G$-map $f: X \to Y$ induces maps $f^H: X^H \to Y^H, f^{(H)}: X^{(H)} \to Y^{(H)}$, and $f^{\underline{(H)}}: X^{\underline{(H)}} \to Y^{\underline{(H)}}$ since for every $x \in X$, we have $G_x \leq G_{f(x)}$.
Observe however that $f$ maps the pair $(X_{(H)}, X_H)$ to $(Y_{(H)}, Y_H)$ only if $G_{f(x)}$ equals $G_x$ for all points $x$ on orbits of type $(G/H)$. This is the case if and only if $f$ is injective on each such orbit $Gx$, i.e. if $Gx$ gets mapped homeomorphically onto $Gf(x)$. If this holds at all points in $X$, we say that $f$ is *isovariant*.

**2.7** By a *$G$-module* $M$, we mean a real vector space of finite dimension equipped with a linear $G$-action from the left. Thus, $M$ is a linear representation of $G$ over $\mathbb{R}$. We will emphasize specifically, when $M$ is to be a complex $G$-module.
As $G$ is a compact group, we can integrate over $G$. Therefore, any linear representation of $G$ is equivalent to an orthogonal or a unitary one. More generally, we can equip any metrizable $G$-space $X$ with a *$G$-invariant metric* which in turn induces a metric on $X/G$ generating the identification topology.

**2.8** The *Tietze-Gleason theorem* is the equivariant version of the Tietze extension theorem: If $X$ is a normal $G$-space, then any $G$-map $f$ from a closed $G$-subspace $A \subset X$ to a $G$-module $M$ can be extended to a $G$-map $F: X \to M$.
In other words, a $G$-module is an *equivariant absolute extensor*, a $G$-AE in short, for the class of normal $G$-spaces.

If the $G$-map $f$ given on $A$ takes its values in some real interval $I$ with trivial $G$-action, then, of course, we can arrange that its $G$-extension $F$ remains in $I$ as well.
Or, if $A_0$ and $A_1$ are disjoint closed $G$-subspaces of $X$, then there exists a $G$-function $\tau: X \to [0, 1]$ separating $A_0$ and $A_1$, i.e. such that $\tau(A_i) = i$ for $i = 0, 1$.

**2.9** We show that paracompact $G$-spaces are *$G$-numerable*, that is numerable in an equivariant sense.

**Proposition.** *Let $X$ be a $G$-space with $G$ a compact group. If $\underline{U}$ is a numerable covering of $X$ by $G$-subspaces, then $\underline{U}$ is numerable by $G$-functions.*

PROOF. Let $\{u, U \in \underline{U}\}$, be a partition of unity subordinate to $\underline{U}$. By integrating over $G$, we make the functions $u$ $G$-functions $u^G: X \to [0, 1]$. The support of $u^G$ is contained in $U$ since $U$ is a $G$-subspace of $X$. It remains to show that the familiy $\{u^G\}$ is a locally finite partition of unity.
$\{u\}$ is a locally finite family and therefore, keeping $x \in X$ fixed, we find, for every $g \in G$, a neighbourhood of $gx$ in which only a finite number of the functions $u$ does not vanish identically. I.e. for every $g \in G$, there are neighbourhoods $V_g \subset G$ of $g$ and $W_g \subset X$ of $x$ such that $u(V_g \cdot W_g) = 0$ holds for all $u$ up to a finite number of exceptions, say $u \in \underline{U} - \underline{U}_g$. Now, take a finite subset of $\{V_g\}$ which covers $G$. Let $W$ be the intersection of the corresponding subset of $\{W_g\}$ and $\underline{U}_x$ that of $\{\underline{U}_g\}$ Then $W$ is a neighbourhood of $x$, $\underline{U} - \underline{U}_x$ is a finite set, and for each $U \in \underline{U}_x$, we have $u(GW) = 0$. On $W$, therefore, $u^G$ vanishes for every $U \in \underline{U}_x$. Hence, the family $\{u^G\}$ is locally finite. And at each point $x \in X$, it sums up to

$$\sum_{\underline{U} - \underline{U}_x} u^G(x) = \int_G \sum_{\underline{U} - \underline{U}_x} u(gx) = \int_G \sum_{\underline{U}} u(gx)$$

which is 1. □

**2.10** For the rest of the section, let $G$ be a *compact Lie group.*

A $G$-space $X$ now enjoys a nice local structure provided only, it is completely regular. For then, any point $x \in X$ lies on a $G_x$-slice. This is the *slice theorem of J.L. Koszul* ([Koszul]). Remember that for $H \leq G$, an *$H$-slice* in $X$ is an $H$-subspace $S \subset X$ for which the multiplication $G \times_H S \to GS \subset X$ is an open $G$-embedding. We call $S$ an *$H$-kernel* if we do not care about whether $GS$ is open in $X$ or not.

**2.11** Among the various consequences of the slice theorem, we are particularly interested in the following results:

*In a completely regular $G$-space $X$, with $G$ a compact Lie group, every orbit is a $G$-neighbourhood retract.* In particular, every point $x \in X$ has a neighbourhood in which all isotropy groups are subconjugate to $G_x$. Thus, $X^{(H)}$ and $X^{\underline{(H)}}$ are *closed* in $X$ for all $H \leq G$. This implies for instance that the orbit space $(G/H)^K/N(K)$ is finite for $K \leq H$.

*If all orbits in $X$ are of type $(G/H)$, then the projection $X \to X/G$ is a locally trivial fibre bundle with fibre $G/H$ and group $W(H)$.* Furthermore, $X$ is then $G$-homeomorphic to the total space of the fibre bundle with fibre $X^H$ associated to the locally trivial $N(H)$-principal bundle $G \to G/N(H)$. Thus, *$X$ is $G$-homeomorphic to $G \times_{N(H)} X^H \approx (G/H) \times_{W(H)} X^H$*, and by restriction, $G$-mappings defined on $X$ correspond one-to-one to $W(H)$-mappings of $X^H$.

**2.12** Providing a compact Lie group, we may further profit both by the *meta theorem of R.S. Palais* and the *embedding theorem of G.D. Mostow*.
The latter purports that a separable, metrizable $G$-space $X$ of finite dimension admits a $G$-embedding into some $G$-module if and only if the number of orbit types occurring on $X$ is finite ([Palais 1] and [Mostow]). We say in short that the *orbit structure* of $X$ has to be finite.
The meta theorem is an induction principle: A statement concerning Lie groups is true for all compact Lie groups if it proves true for a fixed compact Lie group $G$, assuming that it holds for all subgroups of $G$ ([Palais 2]).

**2.13** According to Proposition 1.7.32 in [Palais 2], the dimension of the orbit space of a separable, metrizable $G$-space $X$ is bounded by the dimension of $X$. We want to show that $X$ need not be separable. By the dimension of a space, we mean its Lebesgue covering dimension.

**Theorem.** *For a perfectly normal $G$-space $X$ with $G$ a compact Lie group, the following dimension formulae hold:*

$$\dim(X) = \sup_{(H)} \dim(X_{(H)}),$$
$$\dim(X/G) = \sup_{(H)} \dim(X_{(H)}/G).$$

*In both cases, the supremum is taken over all orbit types $(H)$ on $X$. If, in addition, $X$ is paracompact, we have*

$$\dim(X_{(H)}) = \dim(X_{(H)}/G) + \dim(G/H)$$

*which implies in particular* $\dim(X/G) \leq \dim(X)$.

PROOF. The number of conjugacy classes in $G$ is countable since $G$ is a compact Lie group. Hence, $X$ is the countable union of its $G$-subspaces $X_{(H)}$.

$X_{(H)}$ in turn, being a locally closed subspace of $X$, is an $F_\sigma$-set therein, say $X_{(H)} = \bigcup X^{\nu}_{(H)}$, for $X$ is perfectly normal. On the other hand, employing the sum theorem of Čech, $\dim(X^{\nu}_{(H)}) \leq d$ for all $(H)$ and all $\nu$ implies $\dim(X) \leq d$ ([Nagami], 9-10). Therefore, the dimension of $X$ is the supremum of all $\dim(X^{\nu}_{(H)})$.

The property of being perfectly normal devolves from $X$ upon all of its subspaces ([Engelking], 2.1.6). Open subspaces, for instance, being co-zero sets in $X$, embed into $X \times \mathbb{R}$ as closed subspaces and according to a theorem of K. Morita ([Morita 2]), $X \times \mathbb{R}$ is perfectly normal. Thus, $X_{(H)}$ is perfectly normal and hence, the supremum of $\dim(X^{\nu}_{(H)})$ taken over $\nu$ amounts to $\dim(X_{(H)})$. The same way, we derive $\dim(X/G) = \sup \dim(X_{(H)}/G)$ since $X/G$ is perfectly normal if $X$ is (2.3).

If, in addition, $X$ is paracompact, then $X$ and hence $X/G$ are even hereditarily paracompact according to a theorem of C.H. Dowker. Like above, open subspaces of $X$ turn out to be paracompact since a theorem of E. Michael ensures that $X \times \mathbb{R}$ is paracompact if $X$ is.

If in this case, there is a single orbit type $(H)$ on $X$, then $p: X \to X/G$ is a locally trivial bundle with fibre $G/H$ (2.11). Its base, being paracompact, can be covered by a sequence of open subspaces $U$ over each of which $p$ is trivial ([Husemoller], 3.5.4). $U$ and $p^{-1}(U)$ are $F_\sigma$-sets in the perfectly normal spaces $X/G$ and $X$ respectively. As above, we therefore get $\dim(X/G) = \sup \dim(U)$ and $\dim(X) = \sup \dim(p^{-1}(U))$.

Now, the dimension of $p^{-1}(U) \approx U \times G/H$ equals $\dim(U) + \dim(G/H)$: Employing the sum formula of Čech, this follows from a theorem of K. Morita ([Morita 1]) because $U$ is paracompact and because $G/H$, as a compact $n$-manifold, can be covered by a finite number of solid balls. □

## 3. An Equivariant Transversality Theorem

3.1 Throughout, we assume $G$ to be a *compact Lie group.*

If $X$ is a $G$-space with the property that at every point $x \in X$, there exists a linear $G_x$-module as slice, we say that *G acts locally smoothly* on $X$. In this case, the orbit types meeting a given compact subspace of $X$ are finite in number.

A smooth action of a compact Lie group on a smooth manifold $X$ for example is locally smooth, for at any point $x \in X$, the space normal to the orbit $Gx$ may serve as a slice.

Indeed, $Gx$ is a submanifold of $X$ diffeomorphic to $G/G_x$. Furthermore, $X_{(H)}$ is a smooth $G$-submanifold of $X$ for every $H \leq G$ and the locally trivial fibre bundle $X_{(H)} \to X_{(H)}/G$ is a smooth map.

**3.2** The *equivariant transversality theorem* has been proved in full generality by H. Hauschild in [Hauschild 1]. We only require the following simplified version:

**Theorem.** *Let $V$ be an open $G$-subspace in some $G$-module $E$ and let $f: V \to E$ be a smooth map which avoids the value zero on some closed $G$-subspace $C$ of $V$. Suppose further that for each isotropy subgroup $H \leq G$ on $V$, the number of subgroups of $G$ conjugate to $H$ is finite. Then there exists an arbitrarily small $G$-homotopy relative to $C$ connecting $f$ to a smooth $G$-map $\tilde{f}$ which has only regular zeros. On the zero-set of $\tilde{f}$, each orbit $G/H$ is of finite length and hence, has a finite $G$-automorphism group $W(H) = N(H)/H$.*

**3.3 Corollary.** *With $V$ and $E$ as in Theorem 3.2, let $f: \overline{V} \to E$ be a continuous $G$-map, smooth on $V$ and zero-free on the boundary of $V$.*
*Then there exists a continuous $G$-map $\tilde{f}: \overline{V} \to E$, arbitrarily close to $f$, which coincides with $f$ on the boundary of $V$, is smooth on $V$, and which has only regular zeros.*

PROOF. $\overline{V} - V$ can be separated from $f^{-1}(0)$ by open $G$-neighbourhoods. So, let $W$ be an open $G$-neighbourhood of $f^{-1}(0)$ such that $\overline{W} \subset V$.
Then, $C := V - W$ is closed in $V$ and misses $f^{-1}(0)$. Theorem 3.2 provides a $G$-map $f': V \to E$, transversal to zero and arbitrarily close to $f$ which, on $C$, coincides with $f$. By setting $\tilde{f} \mid V := f'$, and $\tilde{f} \mid (\overline{V} - \overline{W}) := f$, we get a continuous $G$-map $\tilde{f}: \overline{V} \to E$ with the required properties. □

**3.4** PROOF OF THEOREM 3.2. We prove the equivariant transversality theorem in four steps by induction on the number of orbit types on $V$. Observe first, that this number is finite because $E$ is a $G$-module (3.1). So, let $H$ be maximal under the isotropy subgroups of $G$ on $V$. Then $V^H$ equals $V_H$ and $V^{(H)} = V_{(H)}$ is a closed $G$-submanifold of $V$ (2.9 and 3.1).

STEP 1. *The theorem is true for the $W(H)$-map $f^H$ in place of $f$.*

A map $V^H \to E^H$ has none but regular zeros if and only if its mapping graph comes transversally to the zero-section $V^H \times \{0\}$ in $V^H \times E^H$. Therefore, we consider the map $(\mathrm{id}, f^H): V^H \to V^H \times E^H$:

$$\begin{array}{ccc} V^H & \xrightarrow{(\mathrm{id}, f^H)} & V^H \times E^H \\ \big\downarrow \eta & & \big\downarrow \xi \\ V^H/W(H) & \xrightarrow{(\mathrm{id}, f^H)^*} & V^H \times_{W(H)} E^H \end{array}$$

$(\mathrm{id}, f^H)$ is a morphism of smooth $W(H)$-principal bundles over $(\mathrm{id}, f^H)^*$ since $W(H)$ acts freely on $V^H$. $V^H \times_{W(H)} E^H$ is the total space of the vector bundle with fibre $E^H$ associated to $\eta$.

According to René Thom's transversality theorem, there exists a smooth homotopy $F^*$ relative to $C^H/W(H)$ between $(\mathrm{id}, f^H)^*$ and a map which is transversal to the zero-section in $V^H \times_{W(H)} E^H$ since $(\mathrm{id}, f^H)^*$ does not vanish there. The composition of the bundle isomorphism $\eta \times \mathrm{id}_I \cong (F^*_0)^*(\xi) \times \mathrm{id}_I \cong (F^*)^*(\xi)$ with the projection $(F^*)^*(\xi) \to \xi$ is a $W(H)$-homotopy $F$ of $(\mathrm{id}, f^H)$ over $F^*$. It is constant on $C^H$ since the isomorphism of $(F^*_0)^*(\xi) \times \mathrm{id}_I$ with $(F^*)^*(\xi)$ is the identity over $C^H/W(H) \times I$ where $F^*$ is constant. Of course, $I$ denotes the unit interval $[0, 1]$.

The homotopy $F$ is smooth because locally, it agrees with $\mathrm{id}_{W(H)} \times F^*$ up to diffeomorphism. This further implies that $F_1$ comes transversally to the zero-section $V^H \times \{0\}$ in $V^H \times E^H$.

Projecting $F$ onto $E^H$ results in a smooth $W(H)$-homotopy relative to $C^H$ from $f^H$ to a map transversal to zero: For, with proj denoting the projection $V^H \times E^H \to E^H$, we get $\mathrm{proj} \circ T_x F_1$ for the derivate of $\mathrm{proj} \circ F_1$ at some point $x \in V^H$ and by virtue of the transversality of $F_1$, the derivate's image is the whole of $E^H$ if $x$ is a zero of $\mathrm{proj} \circ F_1$. □

STEP 2. *We may assume that for every $K \in (H)$, $f^K$ comes transversally to zero.*

For this, we choose an open $G$-neighbourhood $K$ of $C$ in $V$ such that $f$ vanishes nowhere on $\overline{U}$, and apply Step 1 to $f^H$ with $\overline{U}$ instead of $C$: The resulting smooth $W(H)$-homotopy joining $f^H$ to a 0-transversal map $(f^H)^\sim$ provides a smooth $G$-homotopy relative to $\overline{U}_{(H)}$ in $E$, say $\varphi: V_{(H)} \times I \to E$, since $V_{(H)}$ equals $G \times_{N(H)} V_H \approx G/H \times_{W(H)} V^H$. Modulating with a smooth transformation of $I$ which takes the value 0 on $[0, 1/3]$ and 1 on $[2/3, 1]$, we achieve that $\varphi$ becomes a technical homotopy, i.e. that $\varphi_t$ is constant for $t \leq 1/3$ and $t \geq 2/3$.

Since $V_{(H)}$ is a closed $G$-submanifold of $V$, there exist an open tubular $G$-neighbourhood $N$ of $V_{(H)}$ in $V$ and a smooth $G$-deformation $d$ of $N$ onto $V_{(H)}$ relative to $V_{(H)}$. Again, we let $d$ be a technical homotopy, providing that for $t \geq 2/3$, $d_t$ is a $G$-retraction $r$ from $N$ onto

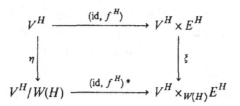

$V_{(H)}$. The two technical homotopies $fd$ and $\varphi r$ combine to a smooth $G$-homotopy $\phi'$ on $N$ which connects $f|N$ to $\varphi_1 r$ and which is constant on $\overline{U}_{(H)}$. We then choose a smooth $G$-function $\tau\colon V \to [0, 1]$ with $\tau(V_{(H)} - U_{(H)}) = 1$ and $\tau(C \cup (V-N)) = 0$ such that $\tau^{-1}(0)$ still is a neighbourhood of $V-N$, and we define $\phi\colon V \times I \to E$ by $(x, t) \mapsto \phi'(x, \tau(x)t)$ for $x \in N$ and $(x, t) \mapsto f(x)$ else.

$\phi$ is a $G$-homotopy of $f$ and $\phi$ is smooth because $\tau^{-1}(0)$ is a neighbourhood of $V-N$. For $x \in C$, we have $\phi(x, t) = f(x)$ and on $V_{(H)}$, $\phi(x, 1) = \phi'(x, \tau(x))$ equals $f(x)$ for $x \in U$ and $\varphi_1 r(x)$ for $x \notin U$ and thus $\varphi_1(x)$ in both cases.

Therefore $\tilde{f} := \phi_1$ is connected to $f$ through a smooth $G$-homotopy which is stationary on $C$, and $\tilde{f}^H = (f^H)^\sim$ is transversal to zero. For any $K = g^{-1}Hg$ in $(H)$, we have $g\tilde{f}^K = \tilde{f}^H g$:

$$\begin{array}{ccc} V^H & \xrightarrow{\tilde{f}^H} & E^H \\ {\scriptstyle g\cdot}\uparrow & & \uparrow{\scriptstyle g\cdot} \\ V^K & \xrightarrow{\tilde{f}^K} & E^K \end{array}$$

Hence, for the derivative at a point $x \in V^K$, we get

$$T_x(\tilde{f}^K)g^{-1} = T_{gx}(\tilde{f}^K g^{-1}) = T_{gx}(g^{-1}\tilde{f}^H) = g^{-1}T_{gx}(\tilde{f}^H).$$

Therefore, $\tilde{f}^K$ is transversal to zero for all $K \in (H)$. □

Let $M$ denote the zero-set of $f$. By now, we may assume that for every $K \in (H)$, $M^K$ is a closed submanifold of $V^K$ of dimension 0.

The normal bundle $M^H \times E$ of the embedding of $M^H$ into $V$ is $N(H)$-homeomorphic to an invariant tubular neighbourhood $\nu$ of $M^H$ in $V$ and $\mu := G\nu$ is a $G$-neighbourhood of $GM^H = M_{(H)}$ in $V$.

Since $G/N(H)$ is finite by assumption, we may assume that, apart from $V^H$, $\nu$ does not meet any of the disjoint closed subspaces $V^K = V_K$, $K \in (H)$, of $V$. Hence, $\mu^H$ coincides with $\nu^H$ and shrinking $\mu$ further, $\mu = G\nu$ splits into $|G/N(H)|$ disjoint copies $\nu(K)$, $K \in (H)$, of $\nu$. For every $K \in (H)$, $\nu(K)$ is then an invariant tubular neighbourhood of $M^K$ in $V$ which is $N(K)$-homeomorphic to $M^K \times E$. Thus, $\mu = G\nu$ is $G$-homeomorphic to $G \times_{N(H)} \nu$.

STEP 3. *For $\nu$ small enough, we may assume beyond Step 2 that, for every $K \in (H)$, $f$ is linear on $\nu(K)$ and that, on the orthogonal complement ${}^{\perp}\nu(K)^K \approx M^K \times {}^{\perp}E^K$, $f$ is the projection onto ${}^{\perp}E^K \subset E$.*

We choose $\nu$ to be contained in $V - C$. Then the saturation $\mu = G\nu$ also lies outside $C$. Let $q$ be the projection from $\nu \approx M^H \times E$ onto $M^H$.
Taking $x \mapsto (T_{q(x)}f)(x - q(x))$ on $\nu^H$ and the projection $M^H \times {}^{\perp}E^H \to {}^{\perp}E^H \subset E$ on ${}^{\perp}\nu^H$, we construct a linear $N(H)$-map $\nu \to E$. Because of the $G$-homeomorphism $G\nu \approx G \times_{N(H)} \nu$, it provides us with a smooth $G$-map $\dot{f}: \mu \to E$. $\dot{f}$ vanishes on $M_{(H)}$ and for every $K \in (H)$, $\dot{f}^K$ is transversal to zero in the points of $M^K$: For, $\dot{f}^K$ is conjugate to $\dot{f}^H$ and the derivate of $\dot{f}^H$ at a point $x \in M^H$ takes the form $(T_x\dot{f})^H = T_x(\dot{f}^H)$ since $\nu^H$ is open in $V^H$ and $q$ is locally constant. $f^H$ however is transversal to zero according to Step 2.
By integrating a suitable smooth function over $G$, we obtain a smooth $G$-function $\rho: V \to I = [0, 1]$ which takes the values 0 on a neighbourhood of $V - \mu$ and 1 on a neighbourhood of $M_{(H)}$. Thus,

$$(x, t) \mapsto (1 - t\rho(x))f(x) + t\rho(x)\dot{f}(x) \text{ for } x \in \mu \quad \text{and} \quad (x, t) \mapsto f(x) \text{ for } x \in V - \mu$$

is a smooth $G$-homotopy $V \times I \to E$. It connects $f$ relative to $C \subset (V - \mu)$ to a G-map $\tilde{f}: V \to E$ which coincides with $\dot{f}$ on the G-neighbourhood $\rho^{-1}(1)$ of $M_{(H)}$.
Since for a sufficiently small invariant tubular neighbourhood $\nu' \subset \nu$ of $M^H$, $G\nu'$ is contained in $\rho^{-1}(1)$, it merely remains to take care that in $V_{(H)}$, $\tilde{f}$ does not vanish anywhere outside $M_{(H)}$. For this, we only have to choose $\nu$ so small right from the beginning that at each instant $t$, the linear homotopy leading $\psi_t$ from $f$ to $\dot{f}$ vanishes on $\mu^H$ only in the points of $M^H$:
On $\mu^H = \nu^H$, we have $\psi_t(x) = (1 - t)f(x) + t(T_{q(x)}f)(x - q(x))$. The map $\Psi := (\psi, \mathrm{id})$: $\mu^H \times I \to E \times I$ has maximal rank in the points of $M^H \times I$ because $q$ is locally constant and $f^H$ is 0-transversal. Hence, there exists an $N(H)$-neighbourhood $U$ of $M^H$ in $\nu$ such that $\Psi$ is injective on $U^H \times I$. In particular, $\psi_t(x)$ vanishes there only for $x \in M^H$. We are done because $U$ of course contains a tubular $N(H)$-neighbourhood of $M^H$ in $V$. ◻

A $G$-map $f$ like in Step 3 with the property that $f^K$ is transversal to zero for every $K \in (H)$, comes transversally to zero in all points of $V_{(H)}$ since at a point $x \in M^K \subset M_{(H)}$, we have

$$(T_xf)({}^{\perp}E^K) = {}^{\perp}E^K \text{ and } (T_xf)(E^K) = \mathrm{im}\big(T_x(f^K)\big) = E^K.$$

Furthermore, on the $G$-neighbourhood $Z := (V - M) \cup \mu$ of $V_{(H)}$ in $V$, $f$ does not vanish since all zeros $f$ has in $v$ lie within $v^H$.

STEP 4. *Proof of the theorem by induction on the number of orbit types on $V$.*

By now, we are done if $V$ equals $V_{(H)}$ and otherwise, we may assume that on a $G$-neighbourhood $Z$ of $V_{(H)}$ in $V$, $f$ has only regular zeros all lying in $V_{(H)}$. Let $X$ and $Y$ be open $G$-neighbourhoods of $V_{(H)}$ in $V$ such that

$$V_{(H)} \subset X \subset \overline{X} \subset Y \subset \overline{Y} \subset Z \subset V.$$

We consider the restriction $f'$ of $f$ to the open $G$-subspace $V' := V - \overline{X}$ of $E$. On $C \cap V'$ and on $\overline{Y} \cap V'$, $f'$ does not take the value zero and together, these spaces form a closed $G$-subspace $C' \subset V'$. Since there are less orbits in $V'$ than in $V$, we find by induction hypothesis a map $(f')^{\sim}\colon V' \to E$, connected to $f'$ through a smooth $G$-homotopy relative to $C$, which has only regular zeros.
Now, $\tilde{f} \mid V' := (f')^{\sim}$ and $\tilde{f} \mid Y := f$ defines a $G$-map $\tilde{f}\colon V \to E$ since $V' \cap Y$ is contained in $C$. $\tilde{f}$ is smooth and transversal to zero because $V'$ and $Y$ are open in $V$. For the same reasons, $f$ and $\tilde{f}$ are connected relative to $C$ through a smooth $G$-homotopy. It remains to show that this homotopy can be kept arbitrarily small in the equivariant sense.
If in Step 1, we choose the homotopy $F^*$ small enough, the graph of its lifting $F$ to $(\mathrm{id}, f^H)$ stays inside a given $W(H)$-neighbourhood of $(\mathrm{id}, f^H)$ and hence, the second component of $F$ will be $W(H)$-close to $f^H$. In Step 2 therefore, we may assume that $\varphi$ stays arbitrarily $G$-close to $f \mid V_{(H)}$. If we keep the tubular neighbourhood $N$ sufficiently small so that $d$ stays next to the identity on $N$, then the composite $\phi'$ of $fd$ with $\varphi r$ will be $G$-close to $f \mid N$. But then, the modified homotopy $\phi_t = \phi'_{\tau t}$, too, will be $G$-close to $f$ since it is constant outside $N$.
In Step 3, finally, the modified linear $G$-homotopy from $f$ to $\tilde{f}$ which is constant outside $Gv$, only moves $G$-marginally away from $f$ as long as the values of $f$ and $\dot{f}$ on $v$ remain in a sufficiently small $N(H)$-neighbourhood of zero. To ensure this, we simply choose $v$ small enough.
The result now follows, as above, by induction. □

The zero-set of $f$ is a zero-dimensional, i.e. discrete submanifold of $V$. Hence, all orbits therein are of finite length. But since for every isotropy group $G_x$ on $V$, the number $|G/N(G_x)|$ of subgroups conjugate to $G_x$ was assumed to be finite, the orbit $Gx \approx G/G_x$ will be finite if and only if its $G$-automorphism group $W(G_x) = N(G_x)/G_x$ is finite. □

# 4. On Equivariant Fibrations and Cofibrations

**4.1** Let $G$ be a *compact group*, not necessarily a Lie group.

By a *$G$-cofibration over a $G$-space $B$* we mean a cofibration in the category of $G$-spaces over $B$ and by a *$G$-fibration* we mean a Hurewicz fibration in the category of $G$-spaces.
The canonical theorems on cofibrations and fibrations all carry over to the equivariant setting. We list the following two examples which both can be found in [tom Dieck-Kamps-Puppe] ([DKP]).

**4.2 Theorem.** *An inclusion $i: A \to X$ of $G$-spaces over $B$ is a $G$-h-cofibration over $B$ if and only if $A$ has a $G$-halo in $X$ which over $B$, is $G$-contractible to $A$.*
*$i$ is a closed $G$-cofibration over $B$ if and only if it is a $G$-h-fibration over $B$ and $A$ is a $(G$-$)$zero-set in $X$.* □

4.3 By virtue of an equivariant formulation of Lemma 8.3 in [DKP], the *section extension theorem,* one of the central theorems in homotopy theory, carries over. Remember that numerable $G$-coverings are numerable by $G$-functions and that paracompact $G$-spaces are $G$-numerable since $G$ is compact (2.9).

**Theorem.** *A $G$-space $p$ over $B$ has the $G$-section extension property, abbreviated $G$-SEP, if the $G$-SEP holds in $p$ over each member of a $G$-numerable covering of $B$. Conversely, the $G$-SEP devolves from $B$ upon any $G$-numerically open subspace.* □

**4.4** With this, we can show that both the $G$-cofibration and the $G$-fibration property are of $G$-local nature.

**Theorem.** *A $G$-space $p$ over $B$ is a $G$-fibration if it is one over each member in a $G$-numerable covering of $B$.*
*An inclusion $A \to X$ of $G$-spaces over $B$ is a $G$-$(h$-$)$cofibration over $B$ if $(A \cap V) \to V$ is one for each member $V$ in some $G$-numerable covering of $X$.*

PROOF. For fibrations, the proof follows the lines of the proof of Theorem 9.4 in [DKP]. In the case of cofibrations we adopt the idea of [Dold 1]:

We define the sheaf $\Gamma$ over $X$ and the associated espace étalé $p: E \to X$ the same way as done there, i.e. non-equivariantly: So, for every open subpace $U$ of $X$, let $\Gamma(U)$ be the set of solutions on $U$. An element $\varepsilon$ in $E$ is then represented by a homotopy on a neighbourhood $U$ of $p(\varepsilon)$ in $X$. If we let $g\varepsilon$ denote the homotopy $(x, t) \mapsto g\varepsilon(g^{-1}x, t)$ on the neighbourhood $gU$ of $gp(\varepsilon)$, then $E$ becomes a $G$-space and $p$ a $G$-map.
A section of $p$ over an open $G$-subspace $U \subset X$ is a $G$-map exactly if the corresponding local solution in $\Gamma(U)$ is a $G$-homotopy. The $G$-SE theorem now provides a global $G$-section of $p$, i.e. a $G$-homotopy on the whole of $X$ with the required properties. □

# CHAPTER II
# Equivariant Vertical Euclidean Neighbourhood Retracts

We define *G-euclidean neighbourhood retracts over B* , abbreviated $G$-ENR$_B$s, following the non-equivariant paragon given in [Dold 2]. Along the lines of [Ulrich], we show that the *familiar properties* of ENR$_B$s carry over to the equivariant setting without any exception. These are the topics treated in the **first two sections** of the present chapter. Beyond, we will discuss the fibration aspect of $G$-ENR$_B$s.
So, we can characterize $G$-ENR$_B$s by the vertical $G$-extension property $G$-ANE$_B$ where ANE stands for *absolute neighbourhood extensor*. Further, a closed vertical $G$-subspace $q$ of a $G$-ENR$_B$ $p$ over a paracompact base is a $G$-ENR$_B$ if and only if the inclusion $q \to p$ is a *G-cofibration over B*. In particular, to be a $G$-ENR$_B$ is a $G$-local property. Over a paracompact base, a $G$-ENR$_B$ $p$ with $G$ a compact Lie group is a *G-fibration* provided $p$ is a proper map. Arbitrary $G$-ENR$_B$s, however, only enjoy *the G-beginning covering homotopy property,* $G$-BCHP in short: $G$-homotopies on the base can be $G$-lifted just a bit.
We will discuss the characterization of spaces having the $G$-BCHP by a $G$-lifting function on their mapping path space as well as a relative version of the $G$-BCHP. With this, we can show that a vertical $G$-space $p: E \to B$ over a $G$-ENR $B$ has the $G$-BCHP if and only if the graph of $p$ is a vertical $G$-neighbourhood retract in the projection proj: $B \times E \to B$. In fact, $B$ needs only be metrizable and $G$-uniformly locally contractible, abbreviated $G$-ULC. The embedding $p \to$ proj is even a closed $G$-$h$-cofibration over $B$ provided $B \times E$ is a normal space. In vertical $G$-spaces, hence, where $B \times E$ is paracompact and $B$ is $G$-ULC, the $G$-BCHP is a $G$-local property with respect to $E$.
Further, a vertical $G$-space over a $G$-ENR $B$ is a $G$-ENR$_B$ if and only if it has the $G$-BCHP and if its total space is a $G$-ENR. This confirms a *conjecture in* [Dold 3], 1.7., proposed for the non-equivariant case: Over a $G$-ENR $B$, a vertical $G$-space with a separable, metrizable total space of finite orbit structure is a $G$-ENR$_B$ if and only if it has the $G$-BCHP and is locally proper, and if all of its equivariant fibres over the orbits on $B$ are (vertical) $G$-ENRs of uniformly bounded dimension. For this, we have to assume that $G$ is a compact Lie group.

In **Section 3** we take up the suggestion in [Dold 3], 1.1 that an ENR$_B$ is a $C^0$-submersion, i.e. a vertical space equiped with a continuous family of local cross sections.

The notion of a *$G$-$C^0$-submersion* will be defined in three versions of increasing strength. $G$-ENR$_B$s prove to be *simple* $G$-$C^0$-submersions. A proper $G$-ENR$_B$ as well as a $G_B$-locally trivial vertical $G$-space is a *fibred* $G$-$C^0$-submersion. Of course, the term "$G_B$" refers to the family $\{G_b, b \in B\}$ of isotropy subgroups of $G$ on $B$. This strongest version of a $C^0$-submersion has been introduced in [Hurewicz], referred to as a map with *slicing structure.* $G_E$-locally trivial vertical $G$-spaces, finally, are examples for *strong* $G$-$C^0$-submersions. This includes $G$-$C^\infty$-submersions of smooth $G$-manifolds which justifies our notation.
A $G$-$C^0$-submersion turns out to be a vertical $G$-space $p: E \to B$ whose graph is $G_E$-locally a vertical neighbourhood retract in the projection $B \times E \to B$. Over a metrizable $G$-ULC base, therefore, a space having the $G$-BCHP is a simple $G$-$C^0$-submersion, a $G_E$-locally equivariant fibration a strong, and a $G$-fibration a fibred one.
Conversely, a $C^0$-submersion $p: E \to B$ over an arbitrary base $B$ has the BCHP locally with respect to $E$. Hence, $p$ enjoys the BCHP if $B$ is metrizable and ULC, and if $B \times E$ is paracompact. A strong $C^0$-submersion is $E$-locally, and a fibred one $B$-locally a regular fibration. Surjective fibred $G$-$C^0$-submersions over a paracompact base with $G$ a compact Lie group are even $G$-fibrations.
Finally, a vertical $G$-space with a $G$-ENR as total space is a $G$-ENR$_B$ if and only if it is a $G$-$C^0$-submersion. Again, $G$ is provided as a compact Lie group and we have to suppose that the orbit structure of the base is locally finite.
While the $G$-BCHP distinguishes $G$-ENR$_B$s with $G$-ENRs as fibres if the base $B$ is a $G$-ENR, a $G$-ENR$_B$ with a $G$-ENR as total space is characterized as a $G$-$C^0$-submersion. We will see, however, that in a $G$-ENR$_B$ $p$ where $p^H$ is surjective for each $H \leq G$, the base is a $G$-ENR if and only if the total space is one, provided the orbit structure of the base is finite and $G$ is a compact Lie group.

To prove these last results we require a *vertical generalization of the Jaworowski criterion* for $G$-ENRs with a compact Lie group ([Jaworowski]): A closed $G$-subspace $p: E \to B$ of a vertically trivial $G$-ENR$_B$ $B \times M \to B$ is a $G$-ENR$_B$ if and only if for each orbit type $(H)$ outside $E$, $p^H$ is a vertical ENR over $B^H$. For this, we provide the base as a perfectly normal Lindelöf space of locally finite dimension and orbit structure. It suffices to be perfectly normal and paracompact when its dimension and its orbit structure are finite.
The proof will cover the entire **Section 4**. It is based on a *vertical version of the equivariant extension theorem in* [Lashof 1]. But where they provide separable, metrizable spaces, we will content ourselves with spaces having the topological properties mentioned above.

In **Section 5**, we discuss locally trivial fibre bundles with action of a compact Lie group, so-called *$(G, t, A)$-bundles.* $A$ stands for the structure group of the bundle and $t$ is a homomorphism from $G$ to Aut($A$) measuring how much the actions of $G$ and $A$ depart from permuting with each other. For the trivial homomorphism, we write $t = e$.

A $(G, t, A)$-fibre bundle over a compact base, or over a paracompact, compactly generated base of finite dimension is a $G$-ENR$_B$ if its non-equivariant fibre $F$ is an $(A \times_t G)$-ENR. A $G$-$A$-fibre bundle, in particular, i.e. a $(G, e, A)$-fibre bundle with trivial $G$-action on the fibre $F$, is a $G$-ENR$_B$ if $F$ is an $A$-ENR.

Compared to the non-equivariant case where $F$ needs only be an ENR, things are more complicated now since $(G, t, A)$-bundles are not necessarily locally trivial in an equivariant sense. In fact, a locally trivial $(G, t, A)$-fibre bundle of finite type over a paracompact, compactly generated base is a $G$-ENR$_B$ if and only if all its $G$-fibres over the orbits of $B$ are all (vertical) $G$-ENRs.

A locally trivial $(G, t, A)$-bundle $p$ proves to be a fibred $G$-$C^0$-submersion. Hence, $p$ is a $G$-fibration if the base is paracompact. This is the *homotopy theorem in* [tom Dieck 1]. A locally trivial $(G, t, A)$-fibre bundle with a $G$-ENR as total space, in particular, is a $G$-ENR$_B$ if the orbit structure of its base is locally finite.

The $G$-fibres of a locally trivial $G$-$A$-fibre bundle are of the form $G \times_H F \to G/H$ where $H$ acts on $F$ via some homomorphism $H \to A$. Hence, a locally trivial $G$-$A$-fibre bundle of finite type over a reasonable base is a $G$-ENR$_B$ if its non-equivariant fibre $F$ is an $H$-ENR for each of the structure homomorphisms $H \to A$.

A $G$-vector bundle, for instance, is a $G$-$GL(n, \mathbb{R})$-fibre bundle. As such, it is locally trivial if its base is completely regular. Therefore, a $G$-vector bundle of finite type over a reasonable base is a $G$-ENR$_B$.

The projection of a completely regular $G$-space $E$ with one single orbit type $(H)$ onto its orbit space $E/G$ is a locally trivial $(G, e, W(H))$-fibre bundle. It is thus a vertical $G$-ENR if, for example, $E$ is a metric space of finite dimension or a compact space.

Conversely, if $E$ is any $G$-space such that $E \to E/G$ is a vertical $G$-ENR, then there is only one orbit type on $E$ - locally at least. Hence, the projection of a $G$-ENR $E$ onto $E/G$ is a (beginning) $G$-fibration exactly if it is a $G$-ENR$_{E/G}$, i.e. if locally, there is only one orbit type on $E$.

**Section 6** ends the chapter with some *basic examples for* $G$-ENR$_B$s with $G$ a compact Lie group. We first derive an inductive criterion: A vertical $H$-space $q$ where $H$ is any subgroup of $G$, is a vertical $H$-ENR if and only if $\mathrm{id}_G \times_H q$ is a vertical $G$-ENR. From this we deduce that a smooth $G$-manifold of finite orbit structure is a $G$-ENR if its topology has a countable base. Therefore, a $G$-$C^\infty$-submersion of such smooth $G$-manifolds is a $G$-ENR$_B$.

We then show that we get back a $G$-ENR when glueing together two $G$-ENRs $E$ and $E'$ along some common closed $G$-subspace $D$ which is also a $G$-ENR. Actually, we are satisfied with any proper $G$-map $E \supset D \to E'$ in place of the inclusion.
Finally, we prove that the fixed point spaces $p^{(H)}$, $p^{\underline{(H)}}$, and $p_{(H)}$ in a $G$-ENR$_B$ $p\colon E \to B$ over a $G$-trivial base are $G$-ENR$_B$s as well. This provides $p$ with a finite filtration by closed $G$-ENR$_B$s $p_i\colon E_i \to B$ such that $E_i - E_{i-1}$ consists of exactly those points in $E$ which lie on orbits of a fixed type. As an immediate consequence, we obtain a *sum formula* for the Euler-Poincaré characteristic of a compact $G$-ENR $E$, namely

$$\chi(E) = \sum_{(H)} \chi_c(E_{(H)})$$

where the sum runs over all orbit types on $E$. It will serve us as a model for the decomposition of the $G$-fixed point index as presented in Chapter III.

## 1. Definition and General Properties

**1.1** Let $G$ be a *compact group*. We will emphasize explicitly when $G$ is assumed to be a Lie group.

By a *vertical $G$-space (over $B$)* we mean a $G$-map $p\colon E \to B$. A vertical $G$-space of the form $\mathrm{proj}\colon B \times F \to B$ is said to be *vertically trivial.* Products of $G$-spaces always carry the diagonal $G$-action.
A subspace of a $G$-space $X$ is called *$G$-numerically open* when it is the co-zero-set of some $G$-function $X \to [0, 1]$. Since $G$ is compact, these are exactly the numerically open $G$-subspaces of $X$ (see I, 2.2). In perfectly normal $G$-spaces, for instance, all open $G$-subspaces are $G$-numerically open.
Recall that *$G$-modules* are finite-dimensional linear representations of $G$ over $\mathbb{R}$ (I, 2.7). They are denoted by $L$, $M$, or $N$ and $\mathbb{R}^n$ stands for the trivial representation.

**1.2 Definition.** A vertical $G$-space $p\colon E \to B$ is called a $G$-ENR$_B$ or a *vertical* $G$-ENR *(over $B$)* if there exist a $G$-module $M$ and a vertical $G$-embedding of $p$ into the projection

$B \times M \to B$ such that therein, $p$ is a vertical $G$-retract of a $G$-numerically open neighbourhood. "ENR" stands for *euclidean neighbourhood retract.*

The properties of $G$-ENR$_B$s listed below are quite analogous to those of ENR$_B$s found in [Dold 2], Section 1. Therefore, proofs will only be outlined when they seem sufficiently non-trivial or when the equivariant setting necessitates a special argument. We refer the reader to [Ulrich] for detailed proofs of the non-equivariant versions.

1.3 The following examples are canonical:

**Proposition.** *Let $E \to B$ be a $G$-ENR$_B$.*
*For any $G$-map $B' \to B$, the pullback $E \times_B B' \to B'$ is a vertical $G$-ENR over $B'$.*
*If $E' \to B'$ is another vertical $G$-ENR, then $E \times E' \to B \times B'$ is a vertical $G$-ENR over $B \times B'$, and for $B' = B$, $E \times_B E' \to B$ is a $G$-ENR$_B$.*
*Finally, if $B \to C$ is a $G$-ENR$_C$, then the composite $E \to B \to C$ is a $G$-ENR$_C$.* □

1.4 $G$-ENR$_B$s have been defined as numerical neighbourhood retracts in order that topological properties of the base be inherited by the total space. For this, we rely on the following lemma:

**Lemma.** *A $G$-numerically open subspace $\tau^{-1}(0, 1]$ of a $G$-space $X$ is, by virtue of the mapping $x \mapsto (x, 1/\tau(x))$, $G$-homeomorphic to a closed $G$-subspace of $X \times \mathbb{R}$.* □

**1.5 Corollary.** *Any $G$-ENR$_B$ $E \to B$ embeds over $B$ as a closed $G$-subspace into the product of $B$ with some $G$-module.*
*In particular, $E$ is paracompact, perfectly normal, or metrizable if $B$ enjoys the respective property. In all of these cases, $E$ and even the product of $E$ with any euclidean space is a normal space.*

PROOF. Using Lemma 1.4, any $G$-ENR$_B$ can be embedded as stated. The rest follows from investigations of E. Michael and K. Morita (see [Morita 2]) on topological properties of product spaces (cf [Ulrich] III, 1.3). □

1.6 According to the theorem of Tietze-Gleason, a $G$-module is a $G$-AE for the class of normal $G$-spaces (I, 2.8). This readily implies the first part of the following characterization of $G$-ENR$_B$s.

**Corollary**: *Every* $G$-$\mathrm{ENR}_B$ *is a G-absolute neighbourhood extensor over B* - a $G$-$\mathrm{ANE}_B$ in short - *for the class of normal G-spaces over B.*
*Conversely, let p be a* $G$-$\mathrm{ANE}_B$ *for some class* $G$-$\underline{C}_B$ *of vertical G-spaces over B where B itself is to belong to* $G$-$\underline{C}$. *If p embeds over B as a closed G-subspace into some* $G$-$\mathrm{ENR}_B$, *then p is a* $G$-$\mathrm{ENR}_B$ *in the cases where* $\underline{C}$ *stands for the class of metrizable, of perfectly normal, or of paracompact spaces.*

PROOF. Every $G$-$\mathrm{ENR}_B$ can be embedded over $B$ into a vertically trivial $G$-$\mathrm{ENR}_B$ as a closed $G$-subspace (1.5). In the second assertion, therefore, we may assume that the $G$-$\mathrm{ANE}_B$ $p: E \to B$ under consideration is a closed $G$-subspace of some projection $\mathrm{proj}: B \times M \to B$ with $M$ a complex $G$-module.
In all three cases listed, Lemma 1.5 then guarantees a vertical $G$-extension of the identity on $E$ to some $G$-neighbourhood $U$ of $E$ in $B \times M$. Thus, $p$ is a vertical $G$-neighbourhood retract in proj. Since in each case, $B \times M$ is normal, $U$ contains a $G$-numerical neighbourhood of $E$. □

**1.7 Addendum.** *Let* $p: E \to B$ *be a* $G$-$\mathrm{ENR}_B$, *and let A be closed G-subspace of some vertical G-space X over B with the property that* $X \times [0, 1]$ *is normal.*
*Then, according to 1.6, any vertical G-map* $A \to E$ *can be extended to a vertical G-map* $U \to E$ *on some G-neighbourhood* $U \subset X$ *of A.*
*If* $f_0$ *and* $f_1$ *are vertical G-maps* $U \to E$ *which, on A, are connected by a G-homotopy* $f_t$ *over B, then, on some G-neighbourhood* $V \subset U$ *of A, there exists a vertical G-homotopy from* $f_0$ *to* $f_1$ *extending* $f_t$.

Using the equivariant version I, 2.8 of the Tietze extension theorem, the proof can be accomplished as in the non-equivariant case (cf [Ulrich] I, 1.9). □

**1.8** Corollary 1.6 provides the following examples of $G$-$\mathrm{ENR}_B$s which, of course, might as well be deduced directly from the definition.

**Proposition.** *Let p be a* $G$-$\mathrm{ENR}_B$.
*If B is paracompact, then a vertical G-retract of a G-numerically open subspace of p is a* $G$-$\mathrm{ENR}_B$, *so for example a closed vertical G-neighbourhood retract of p.*
*If B is perfectly normal, then every vertical G-neighbourhood retract of p is a* $G$-$\mathrm{ENR}_B$.

PROOF. First, a vertical $G$-retract of a $G$-$\mathrm{ENR}_B$ is, just as the space itself, a $G$-$\mathrm{ANE}_B$ for the class of normal $G$-spaces over $B$. By 1.6 hence, it is a $G$-$\mathrm{ENR}_B$ in case $B$ is

In a $G$-ENR$_B$ over a paracompact or a perfectly normal base, every open $G$-neighbourhood of a closed $G$-subspace contains a $G$-numerically open neighbourhood (1.5). Therefore, we only have to check whether a $G$-numerically open subspace $q$ of a $G$-ENR$_B$ $p: E \to B$ is a $G$-ENR$_B$ when $B$ is paracompact or perfectly normal. This follows from Corollary 1.6: On the one hand in fact, as an open $G$-subspace of $p$, $q$ is like $p$ a $G$-ANE$_B$ for the class of normal $G$-spaces over $B$. And on the other hand, using 1.3 and 1.4, $q$ can be embedded over $B$ as a closed $G$-subspace into a $G$-ENR$_B$, namely into the composite $E \times \mathbb{R} \to E \to B$. □

**1.9 Theorem.** *Let $q: D \to B$ be a $G$-ENR$_B$ over a paracompact or a perfectly normal base. A closed $G$-subspace $p: E \to B$ of $q$ is a $G$-ENR$_B$ if and only if the inclusion $p \to q$ is a $G$-cofibration over $B$.*

PROOF. Using Addendum 1.7, we derive from Proposition 1.8 that $p$ is a $G$-ENR$_B$ if and only if the inclusion $p \to q$ is a vertical $G$-$h$-cofibration. This follows as in the non-equivariant analogon (cf [Ulrich], p. 11) via the local characterization of vertical $G$-$h$-cofibrations by contracting halos (see I, 4.2). Further, $E$ is a $G$-zero-set in $D$ if $p$ is a $G$-ENR$_B$:
$E$ is a vertical $G$-retract of some $G$-neighbourhood $O \subset D$. Having embedded $D$ over $B$ equivariantly into $B \times M$, the retraction takes the form $(b, x) \mapsto (b, \rho(b, x))$ where $\rho: O \to M$ is a $G$-map. Since $E$ is closed in the normal $G$-space $D$, we can find a $G$-function $\tau: E \to [0, 1]$ taking the values 0 on $E$ and 1 on $D - O$. Then, $(b, x) \mapsto \min\{1, \|x - \rho(b, x)\| + \tau(b, x)\}$ is a non-negative $G$-function with zero-set $E$. For now, it is only defined on $O$, but it can be extended to a $G$-map on the whole of $D$ because $\tau$ tends to one when approaching the boundary of $O$. □

**1.10** *From now on, the base $B$ is assumed to be a paracompact $G$-space.*

According to Proposition 1.5, the total space of a $G$-ENR$_B$ is henceforth paracompact and so, all its coverings by open $G$-subspaces will be $G$-numerable (see I, 2.9). Since being a vertical $G$-cofibration is a $G$-local property (I, 4.4), Theorem 1.9 implies

**Proposition.** *Let $p: E \to B$ be a closed $G$-subspace of some $G$-ENR$_B$ $q: D \to B$ over a paracompact base. If in $E$, every point has a $G$-neighbourhood which is a vertical $G$-ENR over $B$, then $p$ is a $G$-ENR$_B$.*

PROOF. Every point $x \in E$ has an open $G$-neighbourhood $X$ in $E$ such that $p: X \to B$ is a $G$-ENR$_B$. Since $D$ is normal, there exist $G$-functions $\tau_x: D \to [0, 1]$, $x \in E$, taking the values 0 on $E - X$ and 1 on $Gx$ (I, 2.8). The G-neighbourhoods $\tau_x^{-1}(0, 1]$ and $E \cap \tau_x^{-1}(0, 1]$ of $x$, which are numerically open in $D$ and $X$ respectively, are both $G$-ENR$_B$s. Therefore, the inclusions $(E \cap \tau_x^{-1}(0, 1]) \to \tau_x^{-1}(0, 1]$ are $G$-cofibrations over $B$ according to Theorem 1.9. Since the subspaces $\tau_x^{-1}(0, 1] \subset D$ together with $D - E$ form an open $G$-covering of $D$, the inclusion $E \to D$ is a $G$-cofibration over $B$, for $D$ is paracompact. Hence, $p: E \to B$ is a $G$-ENR$_B$, again by Theorem 1.9. □

**1.11 Proposition.** *Let $p: E \to B$ be a vertical $G$-space over a paracompact, compactly generated base with $G$ a compact Lie group and suppose $p$ is a finite union of open $G$-subspaces $p_i: E_i \to B$ each of which embeds over $B$ as a closed $G$-subspace into the product of $B$ with some $G$-module $M_i$.*
*Then there exist a $G$-module $M$ and a closed vertical $G$-embedding of $p$ into the projection $B \times M \to B$. If, in particular, each $p_i$ is a $G$-ENR$_B$, then $p$ is a $G$-ENR$_B$* according to Proposition 1.10.

PROOF. By hypothesis, there exists a closed vertical $G$-embedding $H_i: E_i \to B \times M_i$ for each $i$. Let $S_i$ denote the one-point compactification of $M_i$: This is a compact, smooth $G$-manifold whereby $G$ acts trivially on the point $\infty$ at infinity.
We define vertical $G$-maps $h_i: E \to B \times S_i$ by $x \mapsto H_i(x)$ for $x \in E_i$ and $x \mapsto (p(x), \infty)$ else. To check continuity, take some $A$ closed in $S_i$. If $\infty$ is in $A$, then $O := S_i - A$ is an open subspace of $M_i$ and hence, $h_i^{-1}(B \times O) = H_i^{-1}(B \times O)$ is open in $E_i$ and thus in $E$. Otherwise, $A$ is a compact subspace of $M_i$ and for each compactum $C \subset B$, we consider the intersection $p^{-1}(C) \cap h_i^{-1}(B \times A) = H_i^{-1}(C \times A)$. Via $H_i$, it is homeomorphic to $(C \times A) \cap H_i(E_i)$ and this is a compact space, for $H_i$ is a closed embedding. Thus, $h_i^{-1}(B \times A) \subset E$ lies properly over $B$ and whence is closed in $E$. For the latter, we have used that the topology of $B$ is compactly generated (see Appendix, A.4 and A.5).
Now, the fibrewise product of all $h_i$ over $B$ forms an injective vertical $G$-map $h: E \to (\prod (B \times S_i))_B = B \times \prod S_i$. $h$ is an embedding: For an open subspace $V \subset E_1$, for instance, $h(V)$ equals $h(E) \cap (h_1(V) \times \prod_{i>1} S_i) = h(E) \cap (H_1(V) \times \prod_{i>1} S_i)$. This is an open subspace of $h(E) \cap (H_1(E_1) \times \prod_{i>1} S_i) = h(E_1)$ because $H_1$ is an embedding, and $h(E_1) = h(E) \cap (B \times M_1 \times \prod_{i>1} S_i)$ is open in $h(E)$.
The product of all $S_i$, which is a compact smooth $G$-manifold, embeds as a $G$-subspace into some $G$-module $M$ by virtue of Mostow's theorem (see I, 2.12). Combining this embedding with $h$, we get a vertical $G$-embedding of $p: E \to B$ into the projection $B \times M \to B$. $p$ is a locally proper map since each of the open subspaces $E_i \subset E$, being closed

in $B \times M_i$, lies locally proper over $B$. Hence, $E$ is a locally closed subspace of $B \times M$ (see Appendix, A.5) and therefore, it is closed in some $G$-numerically open subspace of $B \times M$. Lemma 1.4 now provides a closed vertical $G$-embedding of $p$ into the projection $B \times M \times \mathbb{R} \to B$. □

**1.12** Submersions of smooth manifolds are $G$-ENR$_B$s as we will see in Section 6. Therefore, the following theorem improves the *fibration theorem of C. Ehresmann.*

**Proposition.** *Let $p: E \to B$ be a $G$-ENR$_B$ over a paracompact base with $G$ a compact Lie group. Then $p$ is a $G$-fibration provided it is a proper map.*

PROOF. Let $B \times M \supset O \xrightarrow{r} E$ be a vertical $G$-neighbourhood retraction in $B \times M \to B$ onto a proper $G$-subspace $p: E \to B$. According to Proposition I, 4.4, it suffices to find a covering of $B$ by open $G$-subspaces such that $p$ is a $G$-fibration over each of these spaces, for $B$ is paracompact by assumption. We fix some point $b \in B$ and denote $G_b$ by $H$.
The fibre $p^{-1}(b)$ is compact since $p$ is a proper map, and hence, there exist an open $H$-neighbourhood $V \subset B$ of $b$ and an open subspace $W \subset M$ such that $V \times W$ is a neighbourhood of $p^{-1}(b)$ contained in $O$. As a neighbourhood of the $H$-subspace $\mathrm{proj}_M(p^{-1}(b))$ of $M$, $W$ may be chosen to be $H$-invariant. Further, being proper, $p$ is a closed map. Therefore, $U := \{v \in V: p^{-1}(v) \subset V \times W\}$ is an open $H$-neighbourhood of $b$ in $V$ with the property $p^{-1}(U) \subset U \times W$. Now, let $S \subset U$ be an $H$-slice in $B$ at the point $b$: That is, $GS$ is a $G$-neighbourhood of $b$, $G$-homeomorphic to $G \times_H S$. Then, $p^{-1}(S)$ is contained in $S \times W \subset O$ and as a fibrewise map, $r$ maps $S \times W$ to $p^{-1}(S)$. In other words, $r$ is an $H$-retraction of $S \times W$ onto $p^{-1}(S)$ over $S$.
Since $p$ is a $G$-map, $p^{-1}(S)$ is an $H$-slice in $E$, i.e. $p^{-1}(GS) = Gp^{-1}(S)$ is $G$-homeomorphic to $G \times_H p^{-1}(S)$. So, over the $G$-neighbourhood $GS$ of $b$, $p$ is $G$-homeomorphic to the vertical $G$-space $\mathrm{id}_G \times_H p : G \times_H p^{-1}(S) \to G \times_H S$ which at last is a $G$-fibration:
For, via the vertical $G$-map $\mathrm{id}_G \times_H r$ over $G \times_H S$, $\mathrm{id}_G \times_H p$ is a vertical $G$-retract of $q: G \times_H (S \times W) \to G \times_H S$, the fibre bundle with fibre $W$ associated to the $H$-principal bundle $G \times S \to G \times_H S$. And $q$ is a $G$-fibration as shown in [Lashof 2], 2.12 which, by the way, we will reprove lateron in Proposition 5.7. □

## 2. An Equivariant Beginning Covering Homotopy Property

**2.1** A $G$-$\mathrm{ENR}_B$ which as a map is proper, is a $G$-fibration whereas in arbitrary $G$-$\mathrm{ENR}_B$s, $G$-homotopies on the base can only be lifted just a bit. Therefore, we propose the following definition generalizing [Dold 3], 1.6:

**Definition.** A vertical $G$-space $p: E \to B$ is said to have the *$G$-beginning covering homotopy property*, $G$-BCHP in short, if for every commutative diagram of $G$-maps

$$\begin{array}{ccc} X \times \{0\} & \xrightarrow{\ f\ } & E \\ \downarrow{\scriptstyle \subset} & & \downarrow{\scriptstyle p} \\ X \times [0,1] & \xrightarrow{\ \varphi\ } & B \end{array}$$

there exists, over $\varphi$, a $G$-extension $\phi$ of $f$ to a $G$-neighbourhood $U \subset X \times [0, 1]$ of $X \times \{0\}$. We call the pair $(\phi, U)$ a *beginning $G$-lift of $\varphi$ to $f$.*

**2.2** EXAMPLE. *A vertical $G$-neighbourhood retract of a space with the $G$-BCHP has the $G$-BCHP, like for instance every $G$-$\mathrm{ENR}_B$.* □

**2.3** REMARK. If $X$ is a *paracompact $G$-space*, then every $G$-neighbourhood of $X \times \{0\}$ in $X \times [0, 1]$ contains a *tubular neighbourhood* of the form $X_\varepsilon = \{(x, t) \in X \times [0, 1]: t \leq \varepsilon(x)\}$ with $\varepsilon: X \to (0, 1]$ a continuous $G$-function. Since $X_\varepsilon$ is a $G$-retract of $X \times [0, 1]$, every $G$-lift beginning on $X_\varepsilon$ can be extended to a $G$-map on the whole of $X \times [0, 1]$.
Therefore, *$p$ has the $G$-BCHP for paracompact $G$-spaces $X$ if and only if for every pair $(f, \varphi)$ of $G$-maps as in 2.1, there exist a $G$-homotopy $\phi$ of $f$ on $X$ and a positive $G$-function $\varepsilon$ such that $p\phi(x, t) = \varphi(x, t)$ holds for $t \leq \varepsilon(x)$.*

PROOF. Given a neighbourhood $U$ of $X \times \{0\}$ in $X \times [0, 1]$, it remains to construct a suitable $G$-function $\varepsilon: X \to (0, 1]$.
For every $x \in X$, $(gx, 0)$ has a neighbourhood of the form $V_g \times [0, \varepsilon_g] \subset X \times [0, 1]$ for each $g \in G$. A finite number of the subspaces $V_g$ suffices to cover $Gx$, those corresponding to $g_1, \ldots, g_n$, say, and this neighbourhood contains an open $G$-neighbourhood $V_x$ of $Gx$. Let $\varepsilon_x$ denote the minimum of the $\varepsilon_{g_i}$, $i = 1, \ldots, n$, and let $\{v_x, x \in X\}$ be a $G$-partition of unity

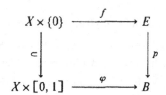

subordinate to the open $G$-covering $\{V_x,\ x \in X\}$ of $X$ (see I, 2.9). Then, $\sum_{x \in X} \varepsilon_x v_x(-)$ is a $G$-function $\varepsilon: X \to [0, 1]$ for which $X_\varepsilon$ lies in $U$. □

**2.4** Let $p: E \to B$ be a vertical $G$-space. For the rest of the section, we let $I$ denote the unit interval $[0, 1]$.
The free path space $B^I$ carries an obvious $G$-structure. A $G$-homotopy in $B$ is just a $G$-map with range in $B^I$ and the *mapping path space of* $p$, $W(p) := \{(x, w) \in E \times B^I : p(x) = w(0)\}$, is a $G$-subspace of the product $E \times B^I$.

**Proposition.** *If $W(p)$ is paracompact, e.g. if $E$ and $B$ are metrizable, then $p$ has the $G$-BCHP for paracompact $G$-spaces if and only if there exists a $G$-map $\Lambda = (\lambda, \rho): W(p) \to E^I \times (0, 1]$ with the properties $\lambda(x, w)(0) = x$, and $p\lambda(x, w)(t) = w(t)$ for $t \leq \rho(x, w)$.*

PROOF. If $p$ has the $G$-BCHP for paracompact $G$-spaces, we consider the following commutative diagram of $G$-maps:

$$\begin{array}{ccc} W(p) \times \{0\} & \xrightarrow{\text{proj}} & E \\ \cap\big\downarrow & & \big\downarrow p \\ W(p) \times I & \xrightarrow{\text{eval}} & B \end{array}$$

By Remark 2.3, a beginning $G$-lift therein provides a $G$-map $\Lambda = (\lambda, \rho): W(p) \to E^I \times (0, 1]$ with the required properties. Conversely, if $\lambda$ is such a map and if

$$\begin{array}{ccc} X \times \{0\} & \xrightarrow{f} & E \\ \cap\big\downarrow & & \big\downarrow p \\ X \times I & \xrightarrow{\varphi} & B \end{array}$$

is a commutative diagram with $H$-maps $\varphi$ and $f$ for some $H \leq G$, then by

$$\varepsilon(x) := \rho(f(x), \varphi(x, -)) \quad \text{and} \quad \phi(x, t) := (f(x), \lambda(x, -))(t)$$

we define a beginning $H$-lift on an $H$-tubular neighbourhood $X_\epsilon$ - without assuming $X$ to be paracompact! □

**2.5** We have proved even more:

**Proposition.** *Let $p$ be a vertical $G$-space with paracompact mapping path space, which has the* $G$-BCHP *for paracompact $G$-spaces - like for example a* $G$-ENR$_B$ *over a metrizable base. Then, for every subgroup $H \leq G$, $p$ has the* $H$-BCHP *for arbitrary $H$-spaces. In particular, $p^H$ has the* BCHP *for arbitrary spaces.*

Concerning the last remark, we observe that the image of an $H$-trivial space $X$ under an $H$-map $X \to Y$ is contained in $Y^H$. □

**2.6** The characterization of $G$-ENR$_B$s by the $G$-BCHP requires the following relative $G$-BCHP version:

**Lemma.** *Let $p: E \to B$ have the* $G$-BCHP, *and suppose that for some pair $A \subset X$ of $G$-spaces, we are given the following commutative diagram of $G$-maps:*

$$\begin{array}{ccc} X \times \{0\} \cup A \times I & \xrightarrow{f \cup h} & E \\ \Big\downarrow{\scriptstyle \subset} & & \Big\downarrow{\scriptstyle p} \\ X \times I & \xrightarrow{\varphi} & B \end{array}$$

*Then there exists a beginning $G$-lift of $\varphi$ to $f$ relative to $h$ on some $G$-neighbourhood of $X \times \{0\} \cup A \times I$, provided $A$ is a $G$-zero-set in $X$ and $h$ is stationary, i.e. $h(a, t) = h(a)$ for all $(a, t) \in A \times I$. When $B$ is metrizable, then the set of stationary points of $\varphi$, for instance, is a $G$-zero-set in $X$.*

*For arbitrary $h$ again, $\varphi$ has a beginning $G$-lift to $f$ relative to $h$ if the inclusion $A \to X$ is a closed $G$-cofibration.*

PROOF. When $B$ is a metric space, then $x \mapsto \operatorname{diam} \varphi(\{x\} \times I)$ is a continuous, non-negative function on $X$ which vanishes exactly on the set of points stationary under $\varphi$.
If $A$ is the zero-set of some $G$-function $\tau: X \to I$, then by

$$(x, t) \mapsto \varphi(x, t/\tau(x)) \text{ for } t \leq \tau(x) \text{ and } (x, t) \mapsto \varphi(x, 1) \text{ for } t \geq \tau(x)$$

we define a $G$-homotopy $\varphi': X \times I \to B$ of $\varphi_0 = pf$ which agrees with $\varphi$ on $A \times I$. Let $(\phi', U')$ be a beginning G-lift of $\varphi'$ to $f$. Then $U := \{(x, t) \in X \times I: (x, \tau(x)t) \in U'\}$ is a $G$-neighbourhood of $X \times \{0\} \cup A \times I$ in $X \times I$, and $\phi(x, t) := \phi'(x, \tau(x)t)$ defines on $U$ a $G$-lift of $\varphi$ to $f$ relative to $h$.

Finally, if $A \to X$ is a closed $G$-cofibration, then the inclusion $X \times \{0\} \cup A \times I \to X \times I$ is one as well according to a theorem of A. Strøm (see [tom Dieck-Kamps-Puppe], 3.20). Therefore, the last remark will follow from the next corollary which, in turn, is a consequence of the part of the lemma already proved. □

**2.7 Corollary.** *Let $p: E \to B$ have the G-BCHP and let*

$$\begin{array}{ccc} A & \xrightarrow{\ f\ } & E \\ {\scriptstyle \subset}\downarrow & & \downarrow{\scriptstyle p} \\ X & \xrightarrow{\ \xi\ } & B \end{array}$$

*be a commutative diagram of G-maps. Suppose $A$ is a (G-)zero-set in $X$ having a G-neighbourhood $V \subset X$ which, in $X$, is G-contractible to $A$ relative to $A$. The inclusion $A \to X$ might for instance be a closed G-fibration.*

*Then, on some G-neighbourhood of $A$ in $X$, there exists a G-extension of $f$ over $\xi$.*

PROOF. Let $\psi: V \times I \to X$ be a $G$-homotopy, stationary on $A$, which leads from a $G$-retraction $r: V \to A$ to the inclusion $V \to X$. Above, we have already proved that the $G$-homotopy $\xi\psi: V \times I \to B$ has a $G$-lift $\phi$ to $fr: V \to E$, defined on some $G$-neighbourhood of $V \times \{0\} \cup A \times I$, which, on $A$, is a stationary map equal to $f$. Then $\phi_1$ is a $G$-extension of $f$ over $\xi$. □

**2.8** We are now ready to characterize vertical $G$-ENRs with a $G$-ENR as base by the $G$-BCHP.

**Theorem.** *A vertical G-space $p: E \to B$ over a G-ENR $B$ is a G-ENR$_B$ if and only if $p$ has the G-BCHP and $E$ is a G-ENR.*

PROOF. Every $G$-ENR$_B$ $p: E \to B$ has the $G$-BCHP (2.2). If, in addition, $B$ is a $G$-ENR, then $B \times M$ is a $G$-ENR for every $G$-module $M$ and hence, so is $E$ as a $G$-neighbourhood retract in such a product (1.8).

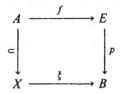

For the converse, we consider the closed $G$-embedding $\gamma := (p, \mathrm{id}_E): E \to B \times E$ over $B$. Since $E$ and $B$ are $G$-ENRs, $B \times E$ is one as well. Hence, by virtue of Theorem 1.9, $\Gamma := \gamma(E)$ has a $G$-neighbourhood $O$ in $B \times E$ which is $G$-contractible to $\Gamma$ relative to $\Gamma$. Let $O \xrightarrow{r} E \xrightarrow{\gamma} \Gamma$ be the corresponding $G$-retraction. Projecting the contraction of $O$ down to $B$, we obtain a $G$-homotopy $\varphi$, stationary on $\Gamma$, which leads from $O \xrightarrow{r} E \xrightarrow{p} B$ to the projection $O \subset B \times E \to B$.

Lemma 2.6 now provides a $G$-lift $\phi$ of $\varphi$ to $r: O \to E$ on some $G$-neighbourhood of $O \times \{0\} \cup \Gamma \times I$ in $O \times I$ which, on $\Gamma$, coincides stationarily with $\gamma^{-1}$. Then $\phi_1$ is a $G$-neighbourhood retraction onto $p$ in the projection $B \times E \to B$ which, at last, is a $G$-ENR$_B$ by Proposition 1.3. □

**2.9** In the following, let $\Gamma = \Gamma_p$ denote the graph of a vertical $G$-space $p: E \to B$ in $B \times E$. It provides a closed vertical $G$-embedding $\gamma = \gamma_p$ of $p$ into the projection proj: $B \times E \to B$. Theorem 2.8 implies in particular that $\gamma$ is a vertical $G$-cofibration if $E$ and $B$ are $G$-ENRs and $p$ has the $G$-BCHP.

This already proves true under weaker assumptions:

**Proposition.** *A vertical $G$-space $p: E \to B$ over a $G$-ULC base has the $G$-BCHP if and only if the graph $\Gamma = \Gamma_p$ of $p$ in the projection* $\mathrm{proj}_B$*: $B \times E \to B$ is a vertical $G$-retract of some $G$-neighbourhood, provided $B$ is metrizable or $\Gamma$ is a zero-set in $B \times E$.*

*In fact, $\Gamma$ has then a $G$-neighbourhood in $B \times E$ which, over $B$, is $G$-contractible to $\Gamma$ relative to $\Gamma$. Therefore, if $B \times E$ is normal, then the graph embedding $\gamma: p \to (B \times E \to B)$ is a vertical $G$-h-cofibration, or, in the case of the zero-set, even a vertical $G$-cofibration.*

**2.10** EXPLANATION. We call a $G$-space $B$ *$G$-uniformly locally contractible,* abbreviated $G$-ULC, if on some $G$-neighbourhood of the diagonal $\Delta$ in $B \times B$, the two projections from $B \times B$ onto $B$ are $G$-homotopic to each other relative to $\Delta$.

From the characterization of $G$-$h$-cofibrations over $B$, we deduce immediately that a $G$-space $B$ whose square $B \times B$ is a normal space, is $G$-ULC if and only if the inclusion $i: \Delta \to B \times B$ is a $G$-$h$-cofibration over $B$. Indeed, $i$ needs only be a $G$-$h$-cofibration, not a vertical one:

For, this means that on some $G$-neighbourhood $O \subset B \times B$ of $\Delta$, there exists a $G$-homotopy $d_t = (\pi_t, \delta_t)$ of the inclusion $O \to B \times B$ relative to $\Delta$ in $B \times B$ which ends with a $G$-retraction onto $\Delta$. Thus, for every point $(x, y) \in O$, the two paths $\pi_t(x, y)$ and $\delta_t(x, y)$ lead from $x$ and $y$ respectively to one and the same point in $B$. Therefore, the composite of $\pi_t(x, y)$ with the reverse path $\bar{\delta}_t(x, y)$ is a path from $x$ to $y$. The family of all these paths constitutes the desired $G$-homotopy connecting the two projections from $O$ onto $B$.

Hence, Proposition 1.9 implies that $G$-ENRs *are* $G$-ULC. More generally, it is easy to see that a $G$-neighbourhood retract in a $G$-ULC space is $G$-ULC as well. Moreover, if $B$ is $G$-ULC, then $B^H$ is ULC for any $H \leq G$.

**2.11** PROOF OF PROPOSITION 2.9. The two projections from $B \times B$ onto $B$ are $G$-homotopic relative to $\Delta$ on a $G$-neighbourhood $O \subset B \times B$ of $\Delta$.
Appending this homotopy to $\mathrm{id}_B \times p : B \times E \to B \times B$, we obtain a $G$-homotopy $\varphi$ in $B$, defined on the $G$-neighbourhood $X := (\mathrm{id}_B \times p)^{-1}(O)$ of $\Gamma$ in $B \times E$, which leads from $p \circ \mathrm{proj}_E$ to $\mathrm{proj}_B$ relative to $(\mathrm{id}_B \times p)^{-1}(\Delta) = \Gamma$. Lemma 2.6 provides a $G$-lift $\phi$ of $\varphi$ to $\mathrm{proj}_E$ on some $G$-neighbourhood $U \subset X \times I$ of $X \times \{0\} \cup \Gamma \times I$ which, on $\Gamma$, coincides stationarily with $\gamma^{-1}$. Hence, if $Y \subset B \times E$ is a $G$-neighbourhood of $\Gamma$ such that $Y \times I$ lies entirely within $U$, then $(\mathrm{proj}_B, \phi): Y \times I \to B \times E$ is a vertical $G$-homotopy of the inclusion $Y \to B \times E$ relative to $\Gamma$ which, at the time $t = 1$, reaches $\Gamma$ because of $p\phi_1(b, x) = \varphi_1(b, x) = b$. □

**2.12** We can now show that over a metrizable $G$-ULC base, the $G$-BCHP is a local property with respect to the total space:

**Corollary.** *Let $p: E \to B$ be a vertical $G$-space over a $G$-ULC base such that $B \times E$ is paracompact, and suppose $B$ is metrizable or the graph $\Gamma$ of $p$ is a zero-set in $B \times E$.*
*If every point $x \in E$ has a $G$-neighbourhood which has the $G$-BCHP over $B$, then the graph-embedding $\gamma: p \to (B \times E \to B)$ is a vertical $G$-h-cofibration and in the case of the zero-set, even a vertical $G$-cofibration. In particular, as a vertical $G$-neighbourhood retract in a vertically trivial $G$-space, $p$ enjoys the $G$-BCHP globally.*

PROOF. Every $x \in E$ has a $G$-neighbourhood $X$ such that $p: X \to B$ has the $G$-BCHP. We may assume that each $X$ is numerically open in $E$. For, $E$ is a normal space as a closed subspace of $B \times E$, and the $G$-BCHP is inherited by open $G$-subspaces.
Then each $B \times X$ embeds into $B \times E \times \mathbb{R}$ as a closed subspace and $B \times E \times \mathbb{R}$ is paracompact according to a theorem in [Michael] because $B \times E$ is so. Hence, $B \times X$ is paracompact, too. Therefore, the inclusion of $B \times X \cap \Gamma = \gamma(X)$ into $B \times X$ is a $G$-$h$-cofibration over $B$ for each $X$ (2.9). And since $B \times E$ is covered by the open $G$-subspaces $B \times X$, $x \in E$, the result follows from Theorem I, 4.4, observing that $B \times E$ is paracompact. □

**2.13** Products of metric spaces with perfectly normal spaces are perfectly normal ([Morita 2]). Thus, Proposition 2.9 and 1.8 immediately imply the following

**Corollary.** *Let $p: E \to B$ be a vertical G-space with B a perfectly normal G-*ULC *base. Then p is a G-*$ENR_B$ *if p has the G-*BCHP *and if E is a G-*ENR. □

In fact, we will see in the next section that in this case, with $G$ being a Lie group, the base is already a $G$-ENR if its orbit structure is finite and if $p^H: E^H \to B^H$ is surjective for each $H \leq G$.

**2.14** The following characterization of vertical $G$-ENRs over a $G$-ENR in terms of the $G$-BCHP confirms a *conjecture made in* [Dold 3], 1.7 for the non-equivariant case.

**Theorem.** *Let $p: E \to B$ be a vertical G-space over a G-*ENR *with a separable, metrizable total space of finite orbit structure and suppose G is a compact Lie group.*
*Then p is a G-*$ENR_B$ *if and only if it has the G-*BCHP, *if it is a locally proper map, and if all its G-fibres $p^{-1}(Q)$ over the orbits Q on B are (vertical) G-*ENRs *of uniformly bounded dimension.*
*Instead of the latter, it suffices to assume that for each $H \leq G$, all the fibres of $p^H$ be* ENRs *and that the dimensions of the non-equivariant fibres $p^{-1}(b)$ of p be uniformly bounded.*

PROOF. First, a $G$-$ENR_B$ $p$ has the $G$-BCHP (2.2), as a map, it is locally proper (see Appendix), and each of its $G$-fibres $p^{-1}(Q)$ is a vertical $G$-ENR over $Q$ (1.3). Since an orbit of $G$, being a homogeneous $G$-space, is a $G$-ENR (5.2), all $G$-fibres $p^{-1}(Q)$ are $G$-ENRs. Their dimensions are uniformly bounded, namely by $\dim(G) + \dim(M)$ when we have embedded $p$ as a closed $G$-subspace into some vertically trivial $G$-$ENR_B$ $B \times M \to B$. For, as a closed subspace, $p^{-1}(Q)$ has at most the dimension of $Q \times M$ which, in turn, is at most $\dim(Q) + \dim(M)$ by virtue of the product theorem of K. Morita (see [Nagami], 26-5).
Second, we convince ourselves, that the alternative assumptions are really weaker: If the dimensions of the $G$-fibres $p^{-1}(Q)$ are uniformly bounded, then so are the dimensions of the non-equivariant fibres $p^{-1}(b)$ since they are closed subspaces of the $G$-fibres. $p_Q: p^{-1}(Q) \to Q$ has the $G$-BCHP just as $p$ and therefore, $p_Q$ is a vertical $G$-ENR exactly if $p^{-1}(Q)$ is a $G$-ENR, according to Theorem 2.8. In this case, $(p_Q)^H$ is clearly a vertical ENR for each $H \leq G$. In particular, all fibres of $p^H$ are ENRs.
To show at last that $p$ is a $G$-$ENR_B$, we have to prove, following Theorem 2.8, that $E$ is a $G$-ENR. Using the characterization of $G$-ENRs with a compact Lie group due to J. Jaworowski (see Section 4), we check whether $E^H$ is an ENR for each $H \leq G$. For this, we use Borsuk's criterion:
By assumption, $E$ is a separable, metrizable space which inherits from $B$ local compactness via the locally proper map $p$. Moreover, we can cover $E$ by a sequence of closed subspaces

$E_i$ on which $p$ is proper and hence closed: For, every point of $E$ lies in a closed neighbourhood on which $p$ is proper, and $E$, being separable and metrizable, is Lindelöf compact. By a theorem of K. Morita (see [Nagami], 21-1), we thus have

$$\dim(E_i) \leq \dim(p(E_i)) + \sup_{b \in B} \dim\left(E_i \cap p^{-1}(b)\right).$$

Each $E_i$ is closed in $E$ and each $p(E_i)$ in $B$. Therefore, the dimensions of the spaces $E_i$ are uniformly bounded by $\dim(B) + \sup_{b \in B} \dim\left(E_i \cap p^{-1}(b)\right)$ because of the assumptions made, and hence, by the sum theorem of Čech (see [Nagami], 9-10), $E$ is of finite dimension. Thus, as a closed subspace of $E$, each $E^H$ is a locally compact, separable, metrizable space of finite dimension.

According to Borsuk's criterion, $E^H$ will be an ENR if, in addition to the above, it turns out to be locally contractible as a pointed space. Now, $p^H: E^H \to B^H$ has the BCHP by corollary 2.5 and its base and all its fibres are locally contractible as pointed spaces since they were all assumed to be ENRs. Thus, the assertion follows from the lemma below which, for fibrations, has been proved in [Allaud-Fadell]. □

**2.15 Lemma.** *Let $p: E \to B$ have the* BCHP *and suppose $B$ is metrizable. If the base and all the fibres of $p$ are locally contractible as pointed spaces, then so is $E$.*

PROOF. Given a neighbourhood $V_0$ of $x \in p^{-1}(b)$ in $E$, we have to find a neighbourhood which, within $V_0$, can be contracted to $x$ keeping $x$ fixed.

By hypothesis, $b$ has a neighbourhood in $B$ which is contractible to $b$ keeping $b$ fixed. Shrinking $V_0$, we may assume thus that $p(V_0)$ can be contracted in $B$ keeping $b$ fixed. Let $\varphi: V_0 \times I \to B$ be $p \times \mathrm{id}_I$ followed by the contraction of $p(V_0)$. Since $p$ has the BCHP and $B$ is metrizable, Lemma 2.6 provides a pointed lift $\phi$ of $\varphi$ to the inclusion $V_0 \to E$ on a neighbourhood $U$ of $V_0 \times \{0\} \cup \{x\} \times I$ in $V_0 \times I$. By assumption, $x$ has a neighbourhood $W_1$ in $W_0 := V_0 \cap p^{-1}(b)$ which, within $W_0$, contracts to $x$ keeping $x$ fixed. We denote the contracting homotopy by $\psi$ and let $V_1$ be a neighbourhood of $x$ in $E$ meeting $p^{-1}(b)$ in $W_1$. Then, $V_2 := \{v \in V_1 : \{v\} \times I \subset U \text{ and } \phi(\{v\} \times I) \subset V_1\}$ is a neighbourhood of $x$ in $E$ and the homotopy

$$V_2 \times I \to V_0, \quad (v, t) \mapsto \phi(v, 2t) \text{ for } t \leq 1/2, \text{ and } (v, t) \mapsto \psi(\phi(v, 1), 2t - 1) \text{ for } t \geq 1/2$$

is a pointed contraction of $V_2$ to $x$ within $V_0$. □

**2.16** COMMENTS. *Let $p: E \to B$ be a vertical space whose base and whose fibres are all* ENRs. *If $p$ is a fibration, then $p$ is an* $ANR_B$ since, by a theorem of S. Ferry, $E$ proves to be an ANR (2.8). And $p$ will be a fibration if, for instance, $p$ is surjective and proper, and strongly regular ([Ferry]). Thus *if in this case, the dimensions of the fibres of $p$ are uniformly bounded, then we have an* $ENR_B$ according to Theorem 2.14. Conversely, a proper surjective fibration turns out to be a strongly regular map if its base is merely locally pathwise connected ([Addis]).

Hereby, a map is called *strongly regular* if neighbouring fibres are $h$-equivalent via arbitrarily small deformations. All spaces in question are assumed to be separable and completely metrizable and "ANR" refers to this class of spaces.

## 3. Equivariant Continuous Submersions

**3.1** We want to detail the suggestion in [Dold 3], 1.1 that vertical ENRs are $C^0$-submersions.

**Definition.** A vertical $G$-space $p: E \to B$ is called a *(simple) $G$-$C^0$-submersion* if, for every point $x \in E$, there exist $G_x$-neighbourhoods $U \subset B$ of $p(x)$ and $V \subset E$ of $x$, and a continuous family of $G_x$-cross sections $s_v$ of $p$ over $U$ with indices $v \in V$ such that $s_v$ leads through $v$, each. In other words, we demand the map $s: U \times V \to p^{-1}(U) \subset E$, $(u, v) \mapsto s_v(u)$ to be continuous and $G_x$-equivariant, satisfying the conditions $ps(u, v) = u$ and $s(p(v), v) = v$. The triple $(s, U, V)$ will be referred to as the *submersion datum of $p$ at the point $x$*.

A $G$-$C^0$-submersion is said to be *strong* if the cross sections $s_v$ may be chosen so as to range completely within $V$. If, at every point $b$ in the image of $p$, we find a $G_b$-neighbourhood $U$ and thereon a continuous family of $G_b$-cross sections, one through each point in $p^{-1}(U)$, then we call $p$ a *fibred* $G$-$C^0$-submersion. In this case, $s$ is a $G_b$-map from $U \times p^{-1}(U)$ to $p^{-1}(U)$.

**3.2** COMMENTS. Since $G_x$ is a subgroup of $G_{p(x)}$ for every $x \in E$, any fibred $G$-$C^0$-submersion is a simple and even a strong one.

In the literature, fibred $C^0$-submersions are known as maps with the *slicing structure property* ([Hurewicz], [Ungar], [Addis]). D. Addis shows in [Addis], 2.6, that a surjective, strongly regular map (see 2.16), whose fibres are locally contractible and completely metrizable, is a strong $C^0$-submersion - without mentioning the term "submersion" explicitly.

3.3 Concerning the topology of $C^0$-submersions, we note:

**Lemma.** *A $C^0$-submersion $p: E \to B$ is an open map. Locally, $B$ is embedded into $E$ as a neighbourhood retract if $p$ is surjective. In this case, $B$ is an* ENR *provided $E$ is one.*

PROOF. Let $b \in B$ belong to the image of some open subspace $O \subset E$. If $(s, U, V)$ is the submersion datum of $p$ at some point $x \in p^{-1}(b) \cap O$, then $s_x^{-1}(O) \subset p(O)$ is a neighbourhood of $b$ in $B$.
If $p$ is surjective, then, for every $b \in B$, there exists some $x \in p^{-1}(b)$. For the datum $(s, U, V)$ of $p$ at the point $x$, we have $ps_x = \mathrm{id}_U$ wherefore $U$ is a retract of $p^{-1}(U)$. If $E$ is now an ENR, then $B$ is one locally. So, by Proposition 1.10, $B$ is an ENR in case $B$ embeds as a closed subspace into some euclidean space, i.e. if $B$ is locally compact, separable and metrizable, and finite-dimensional.
As the image of an ENR under an open map, $B$ is a locally compact space whose topology has a countable base. Therefore, the metrization theorem of Nagata-Smirnov ensures that $B$ is metrizable.
Locally, $B$ embeds into $E \times \mathbb{R}$ as a closed subspace since $B$ is a local neighbourhood retract in $E$. As a Lindelöf-space, $B$ can be covered by a sequence of closed subspaces whose dimensions are bounded by $\dim(E) + 1$. The sum theorem of Čech ([Nagami], 9-10) therefore implies that $B$ is of finite dimension. □

**3.4 Proposition.** *Let $p: E \to B$ be a simple, a strong, or a fibred $G$-$C^0$-submersion. Then any $G$-pull-back of $p$ is a $G$-$C^0$-submersion of the respective type. And $p^H$ is a respective $C^0$-submersion for every $H \leq G$.*

PROOF. Let $(s, U, V)$ be the submersion datum for $p$ at some point $x \in E$. In the case $x \in E^H$, $s$ is an $H$-map because of $H \leq G_x \leq G_{p(x)}$ and $s^H: U^H \times V^H \to E^H$ is a submersion datum for $p^H$ at the point $x$. If the cross sections $s_v$ stay completely within $V$, then $s^H$ ranges in $V^H$ and for $p$ fibred, we still observe $(p^{-1}(U))^H = (p^H)^{-1}(U^H)$.
Let $f: B' \to B$ be a $G$-map and consider any point $(b', x) \in E' := B' \times_B E$. Then $U' := f^{-1}(U)$, $V' := U' \times_B V$, and $s': U' \times V' \to E'$, $(u',(u'', v)) \mapsto (u', s(f(u'), v))$, are the submersion data for $p' := f^*(p)$ at $(b', x)$: Indeed, $s'$ is a $G_{(b', x)}$-map since $G_{(b', x)} = G_{b'} \cap G_x$ is a subgroup of $G_x$.

$s'$ ranges in $V'$ if the image of $s$ is contained in $V$. If $p$ is a fibred $G$-$C^0$-submersion, i.e. if $s$ is a $G_b$-map $U \times p^{-1}(U) \to E$ with $b = p(x) = f(b')$, then $V' = U' \times_B p^{-1}(U)$ equals $(p')^{-1}(U')$, and $s'$ is a $G_{b'}$-map because $G_{b'}$ is a subgroup of $G_{f(b')} = G_b$. □

**3.5** Simple examples for $G$-$C^0$-submersions are locally trivial vertical $G$-spaces:

EXAMPLE. A vertical $G$-space $p{:}E \to B$ is called *$G_B$-locally trivial* if every $b \in B$ has a $G_B$-neighbourhood $U \subset B$ over which, as a vertical $G_B$-space, $p$ is homeomorphic to the projection $U \times p^{-1}(b) \to U$. As usual, $G$ acts diagonally on $U \times p^{-1}(b)$. Of course, the term "$G_B$" refers to the family $\{G_b,\ b \in B\}$.
If $G$ is a Lie group and if $B$ is completely regular, then, at every point $b \in B_H \subset B$, there exists an $H$-slice $S$ such that over $GS \approx G \times_H S$, $p$ is vertically $G$-homeomorphic to the projection $G \times_H (S \times p^{-1}(b)) \to G \times_H S$. In this case, we say $p$ is *$G$-locally trivial.*
In Section 5, we will encounter the adequate notion of equivariant local triviality in form of locally trivial $G$-bundles.

Obviously, *a $G_B$-locally trivial $G$-space over $B$ is a fibred, and a $G_E$-locally one a strong $G$-$C^0$-submersion.*
By the latter, we mean a vertical $G$-space $p{:}\ E \to B$ wherein, for every $x \in E$, there exist $G_x$-neighbourhoods $X$ of $x$ in $E$ and $Y = p(X)$ of $p(x)$ in $B$ such that, as a vertical $G_x$-space, $p_x{:}\ X \to Y$ is trivial.

**3.6** To justify our notation, we show that $C^\infty$-submersions are $C^0$-submersions.

**Proposition.** *A $G$-submersion $p{:}\ E \to B$ of smooth $G$-manifolds is a $G_E$-locally trivial $G$-space over $B$ and hence a strong $G$-$C^0$-submersion. Hereby, we let $G$ be a compact Lie group.*

PROOF. Consider some $x \in E_H \subset E$. There are $H$-modules as coordinate systems around $x$ and $p(x)$ (see [Palais 2]). Locally hence, $p$ is a non-linear map $(L, 0) \to (M, 0)$ of $H$-modules. $T_0(p){:}\ L \to M$ is an $H$-epimorphism by assumption, for the isomorphism $T_0(L) \cong L$ is an $H$-map because $H$ acts linearly on $L$. Thus, $L$ is $H$-isomorphic to the direct sum of the $H$-modules $M$ and $N := \ker(T_0(p))$ and $f{:}\ L \cong M \oplus N \to M \oplus N \cong L$, $x = (y, z) \mapsto (p(x), z)$ is an $H$-map. Since the determinant of $T_0(f)$ is non-zero, $f$ is locally invertible, i.e. an $H$-chart. And the germ of $pf^{-1}$ is represented by the projection $M \oplus N \to M$ which proves the statement. □

**3.7** In the following, we reveal the relationship with $G$-ENR$_B$s.

**Proposition.** *Let* $p\colon E \to B$ *be a vertical* $G$*-neighbourhood retract in some vertically trivial* $G$*-space* proj: $B \times F \to B$, *such as a* $G$-ENR$_B$. *Then* $p$ *is a* $G$-$C^0$*-submersion. It is a fibred one if* $p$ *is a* $G$*-retract of the whole space* proj, *or if* $p$ *is a proper map.*
Therefore, if B is metrizable and $G$-ULC, then every vertical G-space over B having the $G$-BCHP is a $G$-$C^0$*-submersion*. And a $G_E$*-locally* equivariant fibration over B (see 3.5) is a strong, and a G-fibration a fibred $G$-$C^0$*-submersion.*

PROOF. By assumption, there exist an open $G$-neighbourhood $O$ of $E$ in $B \times F$ and a vertical $G$-retraction $r: O \to E$. Let $q$ be the projection $B \times F \to B$.
Every point $x = (p(x), q(x)) \in E$ has a rectangular neighbourhood $U \times W$ in $O$. Since equivariant maps enlarge isotropy groups, we may assume that $U$ and $W$ are $G_x$-invariant. Then $V := E \cap (U \times W)$ is a $G_x$-neighbourhood of $x$ in $E$ and $s := r(\mathrm{id}_U \times q)$: $U \times V \to U \times W \to E$ is the desired $G_x$-family of cross sections.
If $p$ is proper, then, for every $b \in B$, the fibre $p^{-1}(b)$ is compact and hence has a rectangular neighbourhood $U' \times W$ in $O$. We may assume $U'$ and $W$ to be $G_b$-invariant since $q(p^{-1}(b))$ is a $G_b$-subspace of $F$. As a proper map, $p$ is closed, and hence, $U := \{u \in U' : p^{-1}(u) \subset U' \times W\}$ is a $G_b$-neighbourhood of $b$. Then $V = E \cap (U \times W)$ equals $p^{-1}(U)$ and $s$ is a $G_b$-map from $U \times p^{-1}(U)$ to $E$.
That a vertical $G$-space over a metrizable $G$-ULC base is a $G$-$C^0$-submersion if it has the $G$-BCHP, is known from Proposition 2.9: In fact, we have seen in 2.11 that, as a subspace of the projection $B \times E \to B$, $p$ is then a vertical $G$-retract of some $G$-neighbourhood in $(\mathrm{id}_B \times p)^{-1}(O)$ where $O$ is some $G$-neighbourhood of the diagonal in $B \times B$. If $p$ is even a $G$-fibration, then we may take the entire preimage $(\mathrm{id}_B \times p)^{-1}(O)$ as a retracting neighbourhood. Then every point $b \in B$ has a $G_b$-neighbourhood $U$ such that $p^{-1}(U)$ is a vertical $G_b$-retract of the projection $U \times p^{-1}(U) \to U$ - simply choose $U \times U$ to be contained in $O$. Therefore, $p$ is a fibred $G$-$C^0$-submersion.
In a $G_E$-locally equivariant fibration, there exist, for every $x \in E$, $G_x$-neighbourhoods $X$ of $x$ in $E$ and $Y = p(X)$ of $p(x)$ in $B$ such that $p_x\colon X \to Y$ is a $G_x$-fibration. Since, obviously, a $G$-ULC space is $H$-ULC for every $H \leq G$ and since this property devolves upon open $H$-subspaces (2.10), $Y$ is $G_x$-ULC if $B$ is $G$-ULC. Therefore, if $B$ is in addition metrizable, then $p_x$ is a $G_x$-$C^0$-submersion by the part of the proposition already proved. Its datum $s\colon U \times p_x^{-1}(U) \to p_x^{-1}(U) \subset X$ at the point $p(x)$ is, considered for $p$, the datum of a $G$-$C^0$-submersion at the point $x$. □

**3.8** Conversely, any $G$-$C^0$-submersion $p: E \to B$ is $G_E$-locally an equivariant neighbourhood retract in the projection $B \times E \to B$ with respect to the graph embedding $\gamma$ (see 2.9).

**Proposition.** *A vertical $G$-space $p: E \to B$ is a $G$-$C^0$-submersion if and only if, under the graph embedding $\gamma = \gamma_p$, every $x \in E$ has $G_x$-neighbourhoods $V$ in $E$ and $\hat{V}$ in $B \times E$ such that $V$ is a vertical $G_x$-retract of $\hat{V}$.*
*$p$ is a strong $G$-$C^0$-submersion at the point $x$ if and only if $\hat{V}$ can be chosen as a rectangle, i.e. of the form $U \times V$ with $U$ a $(G_x$-$)$neighbourhood of $p(x)$ in $B$.*
*Finally, $p$ is a fibred $G$-$C^0$-submersion if and only if every point $b$ in the image of $p$ has a $G_b$-neighbourhood $U$ over which $p$, with regard to $\gamma$, is a vertical $G_b$-retract of the projection $U \times p^{-1}(U) \to U$.*

PROOF. If $s: U \times V \to p^{-1}(U) \subset E$ is the datum of a $G$-$C^0$-submersion for $p$ at the point $x \in E$, then $V$ is a $G_x$-neighbourhood of $x$ in $E$ and $\hat{V} := s^{-1}(V)$ one of $\gamma(x)$ in $B \times E$, and $s$ is a vertical $G_x$-retraction from $\hat{V}$ onto $V$. $\hat{V}$ equals $U \times V$ in case $p$ is a strong $G$-$C^0$-submersion, and in case $p$ is a fibred one, $s$ is a vertical $G_b$-retraction, with $b = p(x)$, from $\hat{V} = U \times p^{-1}(U)$ onto $V = p^{-1}(U)$.
Conversely, let some $G_x$-neighbourhood $V$ of $x$ in $E$ be a vertical $G_x$-retract of some neighbourhood in $B \times E$, with respect to $\gamma$, of course. On $V$ then, $p$ is a $G_x$-$C^0$-submersion by Proposition 3.5 which provides for $p$ the datum of a $G$-$C^0$-submersion at the point $x$. If $V$ is a vertical $G_x$-retract of a rectangular neighbourhood $U \times V' \subset B \times E$, then, clearly, $V$ is contained in $V'$ and the vertical $G_x$-retraction from $U \times V$ onto $V$ is the datum of a strong submersion for $p$ at the point $x$. Finally, if $U$ is $G_b$-invariant with $b = p(x)$ and if $V = p^{-1}(U)$ is a vertical $G_b$-retract of the projection $U \times p^{-1}(U) \to U$, then this identifies $p$ as a fibred $G$-$C^0$-submersion at the point $b \in p(E)$. □

**3.9** Observing $(B \times E)^H = B^H \times E^H$ and $(p^H)^{-1}(U^H) = (p^{-1}(U))^H$, we immediately have the following

**Corollary.** *If $p: E \to B$ is a $G$-$C^0$-submersion, then for every $H \leq G$, $p^H$ has $E^H$-locally the BCHP. $p^H$ is $E^H$-locally a regular fibration in case $p$ is a strong $G$-$C^0$-submersion, and $B^H$-locally in case $p$ is a surjective fibred one.* □

**3.10** From Corollary 2.12, we know that over a metrizable ULC base, the BCHP is an $E$-local property. Since any $H$-fixed point set in a $G$-ULC space is ULC (2.10), we have as an immediate consequence:

**Corollary.** *Let $p: E \to B$ be a vertical G-space over a metrizable G-*ULC *base such that $B \times E$ is paracompact. If p is a G-$C^0$-submersion, then $p^H$ has the* BCHP *for every $H \leq G$. In particular, p is a $C^0$-submersion if and only if it has the* BCHP. □

**3.11** If $p$ is a $G$-fibration, then $p^H$ is a fibration for every $H \leq G$ as can easily be seen by considering a $G$-lifting function for $p$ (cf 2.5). The following result now unmasks the last statement in Corollary 3.9 as a special case of this observation.

**Proposition.** *Let G be a compact Lie group. A surjective fibred G-$C^0$-submersion over a completely regular base B is B-locally a G-fibration, and hence globally in case B is paracompact.*
*Over a metrizable G-*ULC *base, therefore, surjective fibred G-$C^0$-submersions coincide with surjective (regular) G-fibrations.*

PROOF. From Proposition 3.8, we know that in a surjective fibred $G$-$C^0$-submersion $p$, every $b \in B$ has a $G_b$-neighbourhood $U$ over which $p$ is a vertical $G_b$-retract of the projection $U \times p^{-1}(U) \to U$. Taking a $G_b$-slice at the point $b \in U$, we get, as in the last section of Proof 1.12, a $G$-neighbourhood of $b$ over which $p$, as a vertical $G$-retract of a $G$-fibration, is itself a $G$-fibration. □

**3.12** REMARK. Together with Proposition 3.7, the last result reveals *why a $G$-$ENR_B$ which as a map is proper, is a G-fibration.* For this, we did not need any surjectivity assumptions since the image of a proper $C^0$-submersion is both open and closed.
Moreover, in the results 3.7, 3.10, and 3.11, we may demand that the graph of $p$ be a zero-set in $B \times E$ rather than assuming that $B$ be metrizable (2.9).

**3.13** While a $G$-$ENR_B$ with a $G$-ENR as base is characterized by the $G$-BCHP, we can now identify a $G$-$ENR_B$ with a $G$-ENR as total space as a $G$-$C^0$-submersion.
Recall from I, 2.10 that a $G$-space $X$ is said to have finite orbit structure if the number of orbit types occuring on $X$ is finite.

**Theorem.** *Let G be a compact Lie group. A vertical G-space $p: E \to B$ whose total space is a G-*ENR, *is a G-$ENR_B$ if and only if it is a G-$C^0$-submersion - provided the orbit structure of B is locally finite.*

PROOF. We may assume that $p$ is a surjective $G$-$C^0$-submersion. For, as a pull-back of $p$, $p': E \to p(E)$ is a $G$-$C^0$-submersion (3.4) and $p(E)$, being an open $G$-subspace of $B$ (3.3), is a vertical $G$-ENR over $B$ of locally finite orbit structure.
According to Lemma 3.3, $B$ is therefore an ENR on which $G$ acts with locally finite orbit structure. Further, $p$ embeds over $B$ as a closed $G$-subspace into some vertically trivial $G$-ENR$_B$ because $E$ is assumed to be a $G$-ENR. Thus, with regard to the vertical version of the Jaworowski criterion to be proved in the next section, it suffices to show that $p^H: E^H \to B^H$ is a vertical ENR for each $H \le G$. Observe that this is the point where $G$ has to be a Lie group.
Now, $p^H$ is a $C^0$-submersion by Proposition 3.4 and $E^H$ is an ENR since $E$ is a $G$-ENR. Like at the beginning, we may assume that $p^H$ is surjective. Therefore, its base is an ENR as well (3.3). But then, $p^H$ has the BCHP by Corollary 3.10, and hence is a vertical ENR by Proposition 2.8. □

**3.14 Corollary.** *Let $G$ be a compact Lie group. A vertical $G$-space, whose base and total space are $G$-ENRs, is a $G$-ENR$_B$ if and only if it has the $G$-BCHP, or if it is a $G$-$C^0$-submersion.*

**3.15** REMARK. Depending on the point of view, some of the assumptions in this summary are superfluous.
So, the Lie structure of $G$ is only needed to identify a $G$-$C^0$-submersion as a $G$-ENR$_B$. And the total space of a $G$-ENR$_B$ is already a $G$-ENR if its base is one. In the last section of Proof 3.3, we have also shown the converse:

**3.16 Corollary.** *In a $G$-ENR$_B$ $p$ where $p^H$ is surjective for each $H \le G$, the total space is a $G$-ENR if and only if the base is one - provided $G$ is a compact Lie group and the orbit structure of $B$ is finite.*

PROOF. This follows from the Jaworowski criterion for $G$-ENRs with a compact Lie group since for every $H \le G$, $B^H$ is an ENR if $p^H$ is surjective as shown in Proof 3.13. □

# 4. A Vertical Jaworowski Criterion

**4.1** If $p: E \to B$ is a $G$-$\mathrm{ENR}_B$, then, obviously, $p^H: E^H \to B^H$ is a vertical ENR over $B^H$ for each subgroup $H \leq G$. In this section, we want to show that the converse holds as well. Throughout, let $G$ be a *compact Lie group*.

**4.2** First, we generalize the equivariant extension theorem due to R. Lashof ([Lashof 1]) to vertical $G$-spaces. But where he provides separable, metrizable spaces, we will content ourselves with just metrizable spaces or, more generally, with paracompact, perfectly normal ones.

**Theorem.** *Let $X \to B$ be a vertical $G$-space where $X$ is paracompact and perfectly normal, and let $q: A \to B$ denote a closed $G$-subspace of finite-dimensional complement in $X$. Suppose the number of orbit types in $X - A$ is finite.*
*If $p: E \to B$ is a vertical $G$-space such that $p^H$ is a vertical* ENR *over $B^H$ for each orbit type $(H)$ in $X - A$, then any vertical $G$-map $f: q \to p$ allows a vertical $G$-extension to some $G$-neighbourhood of $A$ in $X$.*

PROOF. We prove the theorem in four steps: By induction, the theorem will be reduced to the case of a single orbit type $(H)$ on $X - A$. But then, we have to prove the statement for the $W(H)$-map $f^H$ only. Therefore, we first assume that $G$ acts freely on $X$ and then continue inductively.

STEP 1. *The statement holds when G acts freely on X.*

In this case, $\pi: X \to X/G$ is a locally trivial $G$-principal bundle: For, $G$ is assumed to be a compact Lie group and as a normal space, $X$ is completely regular (I, 2.11). Therefore, a $G$-map from $X$ to $E$ is the same as a cross section of the fibre bundle $\pi_E: E \times_G X \to X/G$ with fibre $E$ associated to $\pi$, and it is a vertical one over $B$ exactly if the corresponding cross section ranges in the orbit space of the $G$-subspace $E \times_B X \subset E \times X$. We think of $\pi$ as a left $G$-principal bundle, i.e. $G$ acts from the left on $X$ and from the right on the fibre $G$. Then, regarding $E$ as a right $G$-space via $z \cdot g := g^{-1} z$, the total space $E \times_G X$ of $\pi_E$ is just the orbit space $(E \times X)/G$ of $E \times X$ under the diagonal left action of $G$.
Thus, given $f: p \to q$, we have to extend a partial cross section of the locally trivial bundle $\pi_E$, ranging in $(E \times_B X)/G \subset (E \times X)/G$, from $A/G$ to a neighbourhood in $X/G$. The base

$X/G$ is paracompact since $X$ is (I, 2.2) and therefore, we can cover it by a sequence of open subspaces $U$ over which $\pi$ and hence $\pi_E$ is trivial ([Husemoller], 3, 5.4). With the extension problem solved for each $U$, we can construct the required extension inductively like in Proof 12.2 in [Steenrod].
Under a trivialization of $\pi$ over $U$, $(\pi_E)^{-1}(U) \cap ((E \times_B X)/G)$ gets identified with $E \times_B U$ where $U$ lies over $B$ as the subspace $\{e\} \times U \subset G \times U$ of $X$. So, we have to extend a cross section of $E \times_B U \to U$, given on a closed subspace $A_U \subset U$, to some neighbourhood in $U$. This in turn requires a vertical extension of the corresponding map from $(q_U: A_U \to B)$ to $(p: E \to B)$. Now, as an $\mathrm{ENR}_B$, $p$ is an $\mathrm{ANE}_B$ for the class of normal spaces over $B$ according to Corollary 1.6 and $U$ is a normal space, for $X/G$ is perfectly normal as well as $X$ and this property is hereditary (cf Proof I, 2.11). ◻

STEP 2. *When there is only one orbit type $(H)$ on $X - A$, then the statement holds with the $W(H)$-map $f^H$ in place of $f$.* As usual, $W(H)$ denotes $N(H)/H$.

Under that assumption, $X^H - A^H$ is the free $W(H)$-space $(X - A)_H$ and consequently, the orbit projection $\pi: X^H - A^H \to (X^H - A^H)/W(H) = (X - A)/G$ is a locally trivial $W(H)$-principal bundle. Its base is paracompact and finite-dimensional: For, being a paracompact, perfectly normal space, $X$ and hence $X/G$ is hereditarily paracompact (cf Proof I, 2.13). And since perfect normality is inherited by subspaces anyway, the dimension of $(X - A)/G$ is at most $n := \dim(X - A)$ according to Theorem I, 2.13.
With $C$ denoting some $n$-universal $W(H)$-space, let $c: X^H - A^H \to C$ be a map classifying $\pi$. For $C$, we may choose a compact metric space, namely the Stiefel-manifold of orthogonal $r$-frames in $\mathbb{R}^{r+n+1}$ when we have embedded $W(H)$ into the orthogonal group $O(r)$. Then, $C \times X^H$ is a free $W(H)$-space because $W(H)$ acts freely on $C$, and since $C$ is metrizable, $C \times X^H$ is like $X$ paracompact and perfectly normal ([Michael]). By Step 1, therefore, the composite of the projection $C \times A^H \to A^H$ with $f^H: A^H \to E^H$ admits over $B^H$ a $W(H)$-extension $\varphi$ to some $W(H)$-neighbourhood $U$ of $C \times A^H$ in $C \times X^H$. At this point, we make use of the assumption that $p^H: E^H \to B^H$ be a vertical ENR. Since $C$ is compact, $U$ contains a rectangle $C \times \bar{V}$ where $\bar{V}$ is a closed $W(H)$-neighbourhood of $A^H$ in $X^H$.
Now, $\varphi(c, \mathrm{incl}): \bar{V} - A^H \to C \times \bar{V} \to E^H$ and $f^H: A^H \to E^H$ define together a vertical $W(H)$-map $F^H: \bar{V} \to E^H$. Its continuity has to be checked at points in $A^H$ only. But every $a \in A^H$ has a neighbourhood $V_a$ in $\bar{V}$ such that, under $\varphi(c, \mathrm{incl})$ and $f^H$ respectively, both $V_a - A^H$ and $V_a \cap A^H$ map into a given neighbourhood of $F^H(a) = f^H(a)$: For, since $\varphi$ takes the constant value $f^H(a)$ on $C \times \{a\}$ and $C$ is compact, $a$ has a neighbourhood $V_a'$ in $\bar{V}$ with the property that $\varphi$ stays arbitrarily close to $f^H(a)$ on the whole of $C \times V_a'$. ◻

STEP 3. *The statement holds when there is only one orbit type $(H)$ on $X - A$.*

Using the vertical $W(H)$-extension $F^H: \bar{V} \to E^H$ of $f^H$ constructed in Step 2, we define $F: G\bar{V} \to E$ by $gx \to gF^H(x)$. This makes sense, both on $A$ since $F^H$ coincides with $f$ on $A^H$, and on $X - A$ since $(X - A)^H$ is contained in $X_H$ wherefore $gx = g'x'$ implies $g^{-1}g' \in N(H)$.
$F$ is continuous: As the action of $G$ on $X$ is a closed map, its restriction $G \times \bar{V} \to G\bar{V}$ is closed, whence identifying, and the latter's composition with $F$ is continuous.
$G\bar{V}$ is a neighbourhood of $GA^H = A^{(H)}$ in $GX^H = X^{(H)}$ and therefore, $A \cup G\bar{V}$ is a neighbourhood of $A^{(H)}$ in $A \cup X^{(H)}$. But this is already the whole of $X$ since outside $A$, there are only orbits of type $(H)$. For the same reason, the boundary of $A$ is contained in $A^{(H)}$ because locally, the conjugacy classes of isotropy subgroups always get smaller (see I, 2.11). So, $A \cup G\bar{V}$ is a $G$-neighbourhood of $A$ in $X$, and the union $f \cup F: A \cup G\bar{V} \to E$ is a vertical $G$-extension of $f$ since both $A$ and $G\bar{V}$ are closed in $X$ and since $f$ and $F$ agree on their intersection. □

STEP 4. *Proof of the theorem for the general case.*

Following Theorem I, 1.2, we arrange the orbit types on $X - A$ in a sequence $(H_1) \succ (H_2) \succ \ldots \succ (H_r)$ such that $(H_j) \geq (H_i)$ implies $j \leq i$. We set $X_i := \bigcup_{\nu \leq i} X_{(H_\nu)}$ and proceed inductively.
When $f: A \to E$ has already been extended over $B$ to a $G$-map on some $G$-neighbourhood $W_{i-1}$ of $A$ in $A \cup X_{i-1}$, we consider the $G$-subspace $W_{i-1}$ of $W_{i-1} \cup X_{(H_i)}$. On its complement, the only orbit type is $(H_i)$, and as the intersection of $W_{i-1} \cup X_{(H_i)}$ with $A \cup \bar{X}_{i-1}$, it is closed. To see this, take some $x \in \bar{X}_{i-1}$. Then one $H_j$ with $j \leq i-1$ is subconjugate to $G_x$ because $x$ has a neighbourhood in which all isotropy groups are subconjugate to $G_x$ (I, 2.11). If $x$ is not in $A$, then $(G_x)$ is some $(H_k)$ which implies $k \leq j \leq i-1$. Whence, $x$ belongs to $X_{i-1}$.
Since $X$ is hereditarily paracompact and perfectly normal (see Step 2) and since the dimension of a perfectly normal space bounds the dimension of any of its subspaces ([Nagami], 11-11 and 7-2), we may apply Step 3 to extend our map over $B$ to a $G$-map on some $G$-neighborhood $W_i$ of $W_{i-1}$ in $W_{i-1} \cup X_{(H_i)}$. And $W_i$ is a $G$-neighbourhood of $A$ in $A \cup X_i$ because $W_{i-1} \cup X_{(H_i)}$ is one. □

**4.3 Proposition.** *Theorem 4.2 continues to hold when, modulo $A$, $X$ is locally of finite orbit structure and of finite dimension - provided $X$ is a perfectly normal Lindelöf-space.*

PROOF. Every point $x \in X$ has a neighbourhood $V_x$ such that the orbit structure as well as the dimension of $V_x - A$ are finite. Since orbits are compact, every orbit $Q$ on $X$ has a

closed $G$-neighbourhood $\overline{V}_Q$ in which the complement of $A$ is of finite orbit structure and of finite dimension. The latter follows from the sum theorem of Čech ([Nagami] 9-10) because the dimension of a perfectly normal space bounds the dimension of any of its subspaces ([Nagami], 11-11 and 7-2). When $X$ is Lindelöf compact, there exists a sequence $\{V_i\}$ among the neighbourhoods $V_Q$ which covers $X$.
Then, $X^i := \bigcup_{\nu \leq i} \overline{V}_\nu$ forms a sequence of closed $G$-subspaces of $X$ whose interiors cover the whole of $X$. By induction, we have to solve the extension problem in each $X^i$ for a vertical $G$-map $f^i$ defined on $A^i := (A \cap X^i) \cup \overline{W}^{i-1}$ where $\overline{W}^{i-1}$ is a closed neighbourhood of $A \cap X^{i-1}$ in $X^{i-1}$:
Now, $A^i$ is a closed $G$-subspace of $X^i$ whose complement, as a $G$-subspace of the finite union $X^i - A = \bigcup_{\nu \leq i} (\overline{V}_\nu - A)$, is of finite orbit structure and, as above, of finite dimension. Observing the remark below, Theorem 4.2 now provides a vertical $G$-extension of $f^i$ to a closed $G$-neighbourhood $\overline{W}^i$ of $A^i$ in $X^i$. □

**4.4** COMMENTS. $G$-spaces carrying a *locally smooth G-action* (see I, 3.1), for instance, are locally of finite orbit structure and of finite dimension. The latter follows from the product theorem for metric spaces ([Nagami], 12-14).
Further, *regular Lindelöf spaces are paracompact.* A Lindelöf space, by the way, is perfectly normal if and only if it is hereditarily Lindelöf compact ([Engelking], 8.3.A(b)).

**4.5** We can now generalize the Jaworowski criterion ([Jaworowski]) to vertical $G$-ENRs.

**Corollary.** *Let $p: E \to B$ be a closed G-subspace of a vertically trivial* $G$-ENR$_B$ $B \times M \to B$. *Suppose dimension and orbit structure of B are locally finite.*
*Then p is a* $G$-ENR$_B$ *if and only if for each orbit type $(H)$ on $B \times M - E$, $p^H$ is a vertical* ENR *over $B^H$ - provided B is a perfectly normal Lindelöf space. If dimension and orbit structure of B are finite, B needs only be paracompact and perfectly normal.*

PROOF. $p$ is a $G$-ENR$_B$ if we can extend the identity on $E$ over $B$ to a $G$-map on some $G$-neighbourhood of $E$ in $B \times M$. Let us check whether the conditions of Theorem 4.3 or 4.2 are satisfied:
Along with $B$, $B \times M$ will be perfectly normal ([Morita 2]) and, respectively, Lindelöf compact or paracompact ([Michael]). According to the product theorem in [Morita 1], $B \times M$ is (locally) finite-dimensional if $B$ is. Hence, also the complement of $E$ is (locally) of finite dimension, for $B \times M$ is perfectly normal. It remains to check whether the orbit structure of $B \times M$ is (locally) finite:

One agrees readily that the orbit types of $G$ on $B \times M$ are $G$-conjugacy classes of those orbit types on $M$ which arise from representatives $H = G_b$ of the orbit types of $G$ on $B$: For, the isotropy subgroup of $G$ at some point $(b, x) \in B \times M$ is $G_b \cap G_x$ and $H_x$ is $H \cap G_x$. Since $G$ acts linearly on $M$, any $H \leq G$ leaves on $M$ a finite number of orbit types only. Therefore, $B \times M$ is of finite orbit structure if - and only if - $B$ is so. □

**4.6 Corollary.** *Let $B$ be a $G$-subspace in some $G$-module. Then a vertical $G$-space $p: E \to B$ is a $G$-ENR$_B$ if and only if the orbit structure of $E$ is finite and $p^H$ is a vertical* ENR *over $B^H$ for each $H \leq G$.*

PROOF. If $p$ is a $G$-ENR$_B$, then $E$ is a $G$-subspace of some $B \times M$ and hence of a $G$-module. Therefore, the orbit structure of $E$ is finite.
Conversely, $p$ is an ENR$_B$ by hypothesis. Hence, $E$ is a (closed) subspace in some $B \times \mathbb{R}^n$, wherefore $E$ is separable and metrizable, and of finite dimension. Since the orbit structure of $E$ is assumed to be finite, $E$ embeds as a $G$-subspace into some $G$-module $M$. This provides a vertical $G$-embedding $i$ of $p$ into the projection $B \times M \to B$. $i$ is locally closed because $p$ is an ENR$_B$. Hence, we may assume that $i$ is a closed embedding (1.4). Then the statement is a consequence of the previous corollary. □

## 5. Equivariant Bundles

**5.1** We want to investigate fibre bundles with the action of a *compact Lie group* $G$.

By a $(G, t, A)$-*principal bundle*, we understand - corresponding to a $(\Gamma, \alpha, G)$-bundle in [tom Dieck 1] - a locally trivial $A$-principal bundle $\tilde{p}: \tilde{E} \to B$ whose bundle projection is an equivariant map of left $G$-spaces. As usual, $A$ acts from the right on $\tilde{E}$. $t$ is a homomorphism from $G$ to Aut($A$) measuring whether the actions of $G$ and $A$ commute with each other: $g(\alpha x) = (gx)t_g(\alpha)$. In other words, the *semi-direct product* $A \times_t G$, equipped with the multiplication $(\alpha, g)\cdot(\alpha', g') := (\alpha t_g(\alpha'), gg')$ acts on $\tilde{E}$ from the left via $(\alpha, g)\cdot\tilde{x} := (g\tilde{x})\alpha$. If $t$ is trivial, i.e. $t_g = \mathrm{id}_A$ for all $g \in G$, we speak of a $(G, e, A)$-*bundle*.
A fibre bundle $p = \tilde{p}[F]: \tilde{E} \times_A F \to B$ associated to $\tilde{p}$ is called a $(G, t, A)$-*fibre bundle*. Besides the structure group $A$, also $G$ may act from the left on the non-equivariant fibre $F$, but as above, the two actions have to commute via $t$: $g(\alpha x) = t_g(\alpha)(gx)$. Then $\tilde{E} \times_A F$ carries a canonical $G$-structure induced by the diagonal action of $G$ on $\tilde{E} \times F$.

When $G$ acts trivially on $F$, then for every $g \in G$, $t_g(\alpha)$ must act just like $\alpha$. This comes true, for instance, if $t$ is trivial: In that case, $G$ acts on $p$ by bundle maps and, following [Lashof 2], we say that $p$ is a *G-A-fibre bundle.*

**5.2** A $(G, t, A)$-bundle $p$ over $B$ is called *locally trivial* if $G$-locally, it arises as a $G$-pull-back of $(G, t, A)$-bundles of the form $q_\rho\colon ((A \times_t G)/\Gamma_\rho) \times_A F \to G/H$.
We call $q_\rho$ a *(G-)constituent fibre of p.* As a set, $\Gamma_\rho$ is the graph of a mapping $\rho\colon H \to A$ which respects units, but which multiplicatively obeys the rule $\rho(hh') = \rho(h)t_h(\rho h')$. Thus, $\rho$ becomes a homomorphism when $t$ is trivial. As a subgroup, $\Gamma_\rho$ acts on $A \times_t G$ from the right whereas $A$ acts thereon from the left.
When the structure group $A$ is a compact Lie group, any $(G, t, A)$-bundle with a completely regular total space is locally trivial ([tom Dieck 1]).

**5.3** If $p$ is a locally trivial $(G, t, A)$-fibre bundle over $B$ and $q$ is a constituent fibre in $p$ over $G/H$, then $H$ needs not occur as an isotropy subgroup on $B$. In general, the constituent fibres are no genuine $G$-fibres of $p$.

**Lemma.** *Let $p\colon E \to B$ be a locally trivial $(G, t, A)$-fibre bundle. If $B$ is completely regular, then the equivariant fibres $p^{-1}(Q) \to Q$ over the orbits of $G$ on $B$ may serve as constituent fibres for $p$.*

PROOF. Consider some $G$-chart for $p$. So, let $U$ be an open $G$-subspace of $B$ over which $p$ is the pull-back of a constituent fibre $q_\rho$ under a $G$-map $r\colon U \to G/H$. Then $S := r^{-1}(H)$ is an $H$-slice in $B$, i.e. $U = GS$ is $G$-homeomorphic to $G \times_H S$.
$S$ is completely regular since $B$ is and therefore, at every point $b \in S$, we find an $H_b$-slice $S_b \subset S$. Then $S_b$ is a $G_b$-slice in $B$. For, since the isotropy subgroups of $G$ and $H$ coincide in the points of $S$, it follows easily that $S_b$ is a $G_b$-kernel in $B$ (I, 2.8). And $GS_b$ is open in $B$ because its complement in $GS$ is closed being the image of the closed subspace $G \times (S - HS_b) \subset G \times S$ under the multiplication $G \times S \to GS$. In particular, $r$ decomposes on $GS_b$ into the retraction $r_b\colon GS_b \approx G \times_{G_b} S_b \to G/G_b$ followed by the projection $G/G_b \to G/H$. Pulled back to $G/G_b$, $q_\rho$ becomes the bundle $q_b\colon ((A \times_t G)/\Gamma_b) \times_A F \to G/G_b$ where $\Gamma_b$ is the graph of the restriction of $\rho\colon H \to A$ to $G_b = H_b$.
Thus, $U = GS$ is covered by open $G$-neighbourhoods $U_b = GS_b$ of the points $b \in S$ in such a way that over each $U_b$, $p$ arises as the pull-back of some constituent fibre $q_b$ via a G-retraction $U_b \to G/G_b$. Hence, $q_b$ is the $G$-fibre of $p$ over the orbit $Gb \subset B$. □

**5.4** It is now an easy consequence of our results on $G$-$C^0$-submersions that locally trivial $(G, t, A)$-bundles over a paracompact base are $G$-fibrations ([tom Dieck 1], Theorem 4):

**Theorem.** *A locally trivial $(G, t, A)$-fibre bundle is a fibred $G$-$C^0$-submersion. In particular, it is a $G$-fibration if it is $G$-numerably locally trivial.*

PROOF. The second statement follows from the Propositions 3.11 and I, 4.4 since in a locally trivial bundle, the projection is a surjective map. For the first statement, it suffices to check whether a constituent fibre $q: ((A \times_t G)/\Gamma) \times_A F \to G/H$ is a fibred $G$-$C^0$-submersion (3.4).
We regard the projection $\pi: G \to G/H$ as an $H$-map with respect to conjugation. As a submersion of smooth $H$-manifolds, it is an $H$-$C^0$- submersion (3.6). Thus, over some $H$-neighbourhood $U$ of the coset $H \in G/H$, there exists a local $H$-cross section $\sigma: U \to G$ of $\pi$ leading through $e \in G$. Now, $\sigma(hu) = h\sigma(u)h^{-1}$ just says that the local trivialization $(u, h) \mapsto \sigma(u)h : U \times H \to \pi^{-1}(U)$ of the bundle $\pi$ over $U$ is an $H$-map with respect to left multiplication.
Then, obviously, $(u, \alpha, g, x) \mapsto \left(\alpha, \sigma(u)(\sigma\pi(g))^{-1}g, x\right)$ induces a vertical $H$-retraction of $U \times q^{-1}(U)$ onto $q^{-1}(U)$ in $((A \times_t G)/\Gamma) \times_A F$. Pushing this datum around, we obtain, at any other point $gH \in G/H$, the datum of a fibred $G$-$C^0$-submersion over the neighbourhood $gU$. □

We note that the datum constructed provides a vertical $H$-homeomorphism $U \times F \approx q^{-1}(U)$ where $H$ acts diagonally on $U \times F$. Thus, *any locally trivial $(G, t, A)$-fibre bundle is $G_B$-locally trivial* in the sense of 3.5.

**5.5** From Theorem 3.13 we deduce at once:

**Corollary.** *A locally trivial $(G, t, A)$-fibre bundle over a base of locally finite orbit structure is a $G$-ENR$_B$ if its total space is a $G$-ENR.* □

**5.6** A constituent fibre, for example, has only one orbit type on its base. Thus, Proposition 1.11 together with Lemma 5.3 imply:

**Corollary.** *A locally trivial $(G, t, A)$-fibre bundle of finite type over a paracompact, compactly generated base is a $G$-ENR$_B$ if and only if all its $G$-fibres over the orbits on $B$ are (vertical) $G$-ENRs.*

5.7 REMARK. Let $p$ be a locally trivial $(G, \iota, A)$-fibre bundle over $B$. As in the non-equivariant case, $p$ is *of finite type* if $B$ is compact, or paracompact and finite-dimensional. When $G$ acts trivially on $B$, each $G$-fibre in $p$ is $G$-homeomorphic to the non-equivariant fibre $F$ with its given $G$-structure. Also when $B$ is a free $G$-space, there is, up to $G$-homeomorphy, only one $G$-fibre in $p$, namely $G \times F$ with the diagonal $G$-action. And this is a $G$-ENR exactly if $F$ is an ENR as can easily be seen (6.2).

5.8 The constituent fibres of a locally trivial $(G, e, A)$-fibre bundle (see 5.1) are of the form ${}_\rho q: G \times_H F \to G/H$. $\rho$ is a homomorphism from $H$ to $A$. In $G \times_H F$, $G$ acts on the first factor only and $F$ is to be regarded as an $H$-space via the embedding of $H$ into $A \times G$ as the graph $\Gamma_\rho$ of $\rho$, i.e. via $h \cdot x := (\rho(h), h)x$. For, $[(\alpha, g)\Gamma_\rho, x] \mapsto [g, (\alpha, g)^{-1}x]$ is a vertical $G$-homeomorphism of the constituent fibre $q_\rho: ((A \times G)/\Gamma_\rho) \times_A F \to G/H$ with ${}_\rho q$. Proposition 6.2 will show that ${}_\rho q$ is a vertical $G$-ENR if and only if $F$ is a $\Gamma_\rho$-ENR. Thus

**Corollary.** *Let $B$ be a paracompact, compactly generated $G$-space.*
*A locally trivial $(G, e, A)$-fibre bundle over $B$ of finite type is a $G$-ENR$_B$ if and only if its non-equivariant fibre $F$ is a $\Gamma_{\rho_b}$-ENR for each of the homomorphisms $\rho_b: G_b \to A$ occuring in a local trivialization. This is ensured if, for instance, $F$ is an $(A \times G)$-ENR.*
*Hence, a locally trivial $G$-$A$-fibre bundle over $B$ is a $G$-ENR$_B$ if and only if its fibre $F$ is a $G_b$-ENR for each $\rho_b$.* □

5.9 A real $G$-vector bundle of dimension $n$ is a $G$-$GL(n, \mathbb{R})$-fibre bundle and a homomorphism $G_b \to GL(n, \mathbb{R})$ is nothing but a $G_b$-module structure on its non-equivariant fibre $\mathbb{R}^n$.

**Proposition.** *$G$-vector bundles over a completely regular base are locally trivial as equivariant bundles. Therefore, a $G$-vector bundle of finite type over a paracompact, compactly generated base is a $G$-ENR$_B$.*

PROOF. Let $p: E \to B$ be a $G$-$GL(n, \mathbb{R})$-bundle over a completely regular base. The true reason why $p$ is locally trivial, is its equivalence with a $G$-$O(n)$-bundle, i.e. with a bundle having a compact Lie group as structure group (see [Lashof 2]). However, we want to derive the assertion ad hoc:
Consider some point $b$ in $B_H \subset B$ and take there a chart $(p, \psi): p^{-1}(U) \to U \times \mathbb{R}^n$ for the vector bundle $p$. We may assume that $U$ is $H$-invariant. Then $h \cdot x := \psi(h(p, \psi)^{-1}(b, x))$ makes $\mathbb{R}^n$ an $H$-module $N$ such that, in the fibre over $b$, $\psi$ becomes an $H$-isomorphism.

$(p,\varphi): p^{-1}(U) \to U \times N$ with $\varphi(z) := \int_H h\psi(h^{-1}z)$ is a map of $H$-vector bundles since $H$ acts linearly and since $\psi$ is a linear map. Over $b$, $\varphi$ coincides with the isomorphism $\psi$ because $\psi$ is there an $H$-map. Therefore, also on fibres close to $p^{-1}(b)$, $\varphi$ is an isomorphism. In other words, $p$ is $G_B$-locally trivial (3.5.).
Now, let $S \subset B$ be a small $H$-slice at the point $b$ over which $p$, as an $H$-vector bundle, is trivial. Then $p^{-1}(S)$ is an $H$-slice in $E$ because $p$ is a $G$-map. Over the $G$-neighbourhood $GS$ of $b$, therefore, $p$ may be regarded as the $G$-vector bundle $G \times_H p^{-1}(S) \to G \times_H S$ which is isomorphic to $G \times_H (S \times N) \to G \times_H S$. And at one glance, the latter is seen to be the pull-back of the constituent fibre $G \times_H N \to G/H$. □

**5.10** As a consequence, the following - not necessarily locally trivial - equivariant fibre bundles turn out to be vertical $G$-ENRs:

**Corollary.** *Let $B$ be a compact base, or a paracompact, finite-dimensional one of compactly generated topology.*
*A $(G,t,A)$-fibre bundle over $B$ is a $G$-ENR$_B$ if its non-equivariant fibre $F$ is an* $(A \times_t G)$-ENR. *In particular, a $G$-$A$-fibre bundle over $B$ with an $A$-ENR as fibre is a $G$-ENR$_B$.*

PROOF. Let $F$ be an $(A \times_t G)$-neighbourhood retract in some $(A \times_t G)$-module $M$. If $\tilde{p}: \tilde{E} \to B$ is a principal $(G,t,A)$-bundle, then over $B$, $\tilde{E} \times_A F$ is a $G$-retract of a neighbourhood in $\tilde{E} \times_A M$. The latter is now a $G$-vector bundle over $B$ because $G$ acts on it by bundle maps, and it is of finite type due to our assumptions on $B$ (see 5.7). □

**5.11** EXAMPLE. *The orbit projection $p: E \to E/G$ on a completely regular $G$-space $E$ with one single orbit type $(H)$ is a locally trivial $(G, e, W(H))$-fibre bundle,* but of course not a $G$-$W(H)$-bundle:
For, $p$ is a locally trivial fibre bundle with fibre $G/H$ and structure group $W(H)$ (I, 2.11). $G$ acts trivially on the corresponding principal bundle $E_H \to E/G$, and on the fibre, the actions of $G$ and $W(H)$ commute because $W(H)$ acts on $G/H$ as the group of $G$-automorphisms.
Since the constituent fibre $G/H$ is a $G$-ENR (see 6.2.), *$p$ is a vertical $G$-ENR if $p$ is of finite type and if $E/G$ is paracompact and compactly generated.* In fact, $E/G$ is paracompact and compactly generated exactly if $E$ is, and $p$ is of finite type if $E$ is a compact space, or a paracompact, perfectly normal space of finite dimension such as a *finite-dimensional metric space* (see 5.7 and I, 2.13).

Furthermore, $p$ embeds as a vertical $G$-subspace into a vertically trivial $G$-ENR over $E/G$ if and only if $E$ *admits an isovariant mapping to some $G$-module.* And a closed embedding of that kind exists if, in addition, $E$ *is paracompact and locally compact.* In this case, $p$ *is a vertical $G$-*ENR *without any other assumptions to be made,* according to criterion 1.10.

**5.12** Conversely, regardless of $G$ being a Lie group, we get:

**Proposition.** *Let $p: E \to E/G$ be the orbit projection of any $G$-space. If $p$ is a $G$-*ENR$_{E/G}$, *then, locally, there is only one orbit type on $E$. In other words, $E$ is then the topological sum of its $G$-subspaces $E_{(H)}$.*

PROOF. Let $(s, U, V)$ be the submersion datum for $p$ at some point $x \in E^H \subset E^{(H)}$. Since $s$ is a $G_x$-map, the cross section $s_x: U \to E$ through $x$ lives in $E^{G_x} \subset E^H$. Therefore, the neighbourhood $U = ps_x(U)$ of $p(x)$ in $E/G$ lies entirely in the image $E^{(H)}/G$ of $E^H$. Thus, $E^{(H)}$ and hence $E_{(H)}$ is open in $E$. □

**5.13** The orbit space of a $G$-ENR is an ENR ([tom Dieck 2], 5.2.5.). Therefore, the results 2.8, 1.12, 5.12, and 5.4 imply:

**Corollary.** *The orbit projection of a $G$-*ENR *$E$ is a $G$-*ENR$_{E/G}$ *if and only if it is a (beginning) $G$-fibration. Locally, there is then only one orbit type on $E$.*
*Conversely, the orbit projection of a paracompact $G$-space with one single orbit type is a $G$-fibration.* □

## 6. Miscellaneous Examples

**6.1** As an application, we list some examples required in Chapter III. Throughout, $G$ is assumed to be a *compact Lie group.*

**6.2** We start with an inductive criterion of basic importance. Let $H$ be any subgroup of $G$.

**Proposition.** *A vertical $H$-space $q$ over $C$ is an $H$-*ENR$_C$ *if and only if* $\mathrm{id}_G \times_H q$ *is a vertical $G$-*ENR *over $G \times_H C$. Also, an $H$-space $F$ is an $H$-*ENR *if and only if $G \times_H F$ is a $G$-*ENR.

PROOF. The slice theorem implies that the coset space $G/H$ is a $G$-ENR since it can be realized as an orbit in some $G$-module. The projection $G\times_H F\to G/H$ is a $G$-$C^0$-submersion (5.4). According to Corollary 3.14, therefore, it is a vertical $G$-ENR exactly if $G\times_H F$ is a $G$-ENR. So, it remains to prove the first statement.
If $q$ is a vertical $H$-neighbourhood retract in a vertically trivial $H$-$\mathrm{ENR}_C$ $C\times N\to C$, then $\mathrm{id}_G\times_H q$ is a vertical $G$-neighbourhood retract in $G\times_H(C\times N)\to G\times_H C$. Now, any $H$-module is a direct factor in some $G$-module, hence in particular an $H$-retract thereof. Therefore, we may assume that $N$ is a $G$-module. Then, $(y,c,x)\mapsto(y,c,gx)$ induces a $G$-homeomorphism $G\times_H(C\times N)\approx(G\times_H C)\times N$ over $G\times_H C$, both equipped with the diagonal $G$-action, and the latter is a vertical $G$-ENR.
Conversely, if the $G$-space $\mathrm{id}_G\times_H q$ over $G\times_H C$ is a vertical $G$-ENR, then it is a vertical $H$-ENR and hence, so is its part over the $H$-subspace $C\subset G\times_H C$. But that part is just $q$. □

**6.3** We are now ready to show that a smooth $G$-manifold with finite orbit structure is a $G$-ENR.

**Proposition.** *Let $E$ be a separable, metrizable $G$-space of finite dimension with locally smooth $G$-action (see I, 3.1). Then $E$ is a $G$-*ENR *if and only if the orbit structure of $E$ is finite.*
*For example, if $E$ is a smooth $G$-manifold of finite orbit structure whose topology has a countable base, such as a compact smooth $G$-manifold, then $E$ is a $G$-*ENR.

PROOF. Since $G$ acts locally smoothly on $E$, every point in $E$ has a $G$-neighbourhood of the form $G\times_H N$ with $N$ an $H$-module. Locally hence, $E$ is a $G$-ENR by Proposition 6.2 and in particular, $E$ is locally compact. This together with the remaining assumptions identifies $E$ as a closed $G$-subspace in some $G$-module (see I, 2.12). Therefore, $E$ is a $G$-ENR according to Proposition 1.10.
Concerning the example, one has to observe that a locally compact space, such as a manifold, is both separable and metrizable if its topology has a countable base, and that a compact space is so if it is locally metrizable ([Nagata], p. 212). □

**6.4** Since a submersion of smooth $G$-manifolds is a $G$-$C^0$-submersion (3.6), Corollary 3.14 together with the last result implies:

**Corollary.** *A $G$-submersion of smooth $G$-manifolds of finite orbit structure is a vertical $G$-*ENR *if the topology of both manifolds has a countable base.* □

**6.5** From the Jaworowski criterion, we deduce:

**Proposition.** *Let $E$ and $E'$ be $G$-ENRs and suppose $f: D \to E'$ is a proper $G$-map where $D \subset E$ is a closed $G$-ENR.*
*Then the join $E \cup_f E'$ with its $G$-structure inherited from $E$ and $E'$ is a $G$-ENR. In particular, $E/D$ is a $G$-ENR in case $D$ is compact.*

PROOF. Since $f$ is a $G$-map, $F := E \cup_f E'$ inherits a $G$-action from $E$ and $E'$. $F$ is of finite orbit structure: For, if $y \in F$ comes from $D$ under the projection $q: E \oplus E' \to E \cup_f E'$, then $G_y$ equals $G_{f(x)}$ for any $x \in D \cap q^{-1}(y)$. With regard to the Jaworowski criterion, it suffices to show that $F^H$ is an ENR for every $H \leq G$. Thus, we have to prove the statement for the non-equivariant case only.
First, $q: E \oplus E' \to F$ is a proper map: For, the fibres of $q$ are compact as $f$ is proper, and $q$ is closed because $q^{-1}(q(X \oplus X'))$ equals $(X \cup f^{-1}(X') \cup f^{-1}(f(X))) \oplus (X' \cup f(X))$. As the image of a metric space under a proper map, $F$ is metrizable ([Engelking], 4.4.15) and according to [Hu], Theorem IV, 1.2, $F$ is consequently an ANR for the class of metrizable spaces. I.e: Under any closed embedding into a metrizable space, $F$ is a retract of some neighbourhood. So it remains to show that $F$ embeds as a closed subspace into some euclidean space, i.e. that $F$ is locally compact, separable and metrizable, and finite-dimensional.
Now, $F$ is separable as $E \oplus E'$ is, and local compactness is in question only at points $q(x)$ with $x \in D$. For, $D$ is closed in $E$ by assumption and $f(D)$ is closed in $E'$ because $f$ is a proper map (see Appendix).
Let $U' \subset E'$ be an open neighbourhood of $f(x)$ with compact closure and let $U \subset E$ be an open neighbourhood of $x$ in $E$ such that $U \cap D = f^{-1}(U')$. We may assume that $\overline{U}$ is compact because $f^{-1}(U')$ is contained in the compactum $f^{-1}(\overline{U'})$. Obviously, $q^{-1}(q(U \oplus U'))$ equals $U \oplus U'$ wherefore $q(U \oplus U')$ is an open, and hence $q(\overline{U} \oplus \overline{U'})$ a compact neighbourhood of $q(x)$.
Since we can cover $E$ by a sequence of compact subspaces $E_i$, $F$ will be finite-dimensional if the dimensions of the closed subspaces $F_i := q(E_i \oplus E')$ are uniformly bounded. This, of course, is due to the sum theorem of Čech. Since $f$ is proper and $D$ is closed, we may assume that $E_i \cap D$ equals $f^{-1}(f(E_i))$. Then we have $F_i = E_i \cup_{f_i} E'$ where $f_i$ is the restriction of $f$ to $D_i = E_i \cap D$.
Dividing out $E'$, we get a surjection $p_i: F_i \to E_i/D_i$. This is a closed map since $D$ is closed in $E$, and its range is a metrizable space, being a quotient of a compact metric space. Therefore, according to a theorem of K. Morita (see [Nagami], 21-1), the dimension of $F_i$ is bounded by that of $E_i/D_i$ plus the supremum of the dimensions of all fibres of $p_i$. Now,

$\dim(E_i/D_i)$ is at most $1+\dim(E)$ since any open covering of $E_i/D_i$ has a refinement of order at most $1+(1+\dim(E_i))$. The only non-trivial fibre of $p_i$, which is $q(D_i \oplus E')$, is a closed subspace of the mapping cylinder $Z_i$ of $f_i : D_i \to E'$. Thus, having embedded $E$ and hence $D_i$ into $\mathbb{R}^n$ and $E'$ into $\mathbb{R}^{n'}$, both of them as closed subspaces, we get a closed embedding $Z_i \to [0,1] \times \mathbb{R}^n \times \mathbb{R}^{n'}$. In fact, $Z_i$ embeds as the subspace of all triples $(t, x, x')$ with $t=0$, $x=0$, $x' \in E'$, or $t>0$, $x/t \in D_i$, $x' = f(x/t)$. Therefore, the dimension of $q(D_i \oplus E')$ is bounded by $(1+n+n')$, independently of $i$.
Altogether, we have shown $\dim(F_i) \leq (2+2n+n')$ for each $i$. □

**6.6** Let us stress a special case of the preceding proposition:

**Corollary.** *The union of two G-ENRs, which are closed subspaces of some common G-space, is a G-ENR if their intersection is one.* □

**6.7 Proposition.** *Let $p$ be a $G$-ENR$_B$ with trivial $G$-action on its base. Then, for all $H \leq G$, the fixed point spaces $p^{(H)}$ and $p^{\underline{(H)}}$ are closed $G$-ENR$_B$s in $p$. In particular, their difference $p_{(H)}$ is a $G$-ENR$_B$.*
*Likewise, $p^H$, $p^{\underline{H}}$, and $p_H$ are ENR$_B$s and the former two are closed in $p$.*

PROOF. We have noticed in I, 2.11 that $p^{(H)}$ and $p^{\underline{(H)}}$ are closed in $p$. Moreover, $p^{\underline{H}}$ is the $H$-fixed point space of $p^{\underline{(H)}}$. Thus we must show that $p^{(H)}$ and $p^{\underline{(H)}}$ are $G$-ENR$_B$s.
$p$ is a vertical $G$-neighbourhood retract in some projection $B \times M \to B$. Hence $p^{(H)}$ is one in $(B \times M)^{(H)} \to B^{(H)}$ which is simply the projection $B \times M^{(H)} \to B$ since $G$ acts trivially on $B$. So it remains to show that any $(H)$-fixed point space of a $G$-ENR is a $G$-ENR. Relying on the meta theorem of R. Palais (see I, 2.12), we assume that the assertion is true for all $K$-ENRs when $K$ is a proper subgroup of $G$.
Clearly, it suffices to consider a $G$-module $M$. We may assume $M^G = 0$: For, since $M$ splits off its invariant part $M^G$ as a $G$-direct summand, $M^{(H)}$ decomposes into the product of $M^G$, which is a linear $G$-ENR, and the $(H)$-fixed point space of a $G$-module whose invariant part is trivial. Furthermore, it suffices to check $G$-locally whether $M^{(H)}$ is a $G$-ENR because $M^{(H)}$ is closed in $M$ (1.10).
Consider any point $x \neq 0$ in $M^{(H)}$. In $M$, $x$ has a $G$-neighbourhood $X$ of the form $G \times_K N$ where $N$ is a $K$-module and $K$ is the isotropy subgroup of $G$ at $x$. $K$ is strictly smaller than $G$ and $H$ is subconjugate to $K$. From $G_{[g,z]} = gK_z g^{-1}$ for $[g,z] \in G \times_K N$, we see that $X^{(H)}$ is the union of the $G$-subspaces $G \times_K N^{(g^{-1}Hg)}$, taken over the set of all $g \in G$ which conjugate $H$ into $K$. Modulo $K$, the latter is the fixed point space $(G/K)^H$, and since we

conjugate $H$ with each representative $g$, we only have to take into account representatives of the orbit space $(G/K)^H/N(H)$. And this is now a finite set (see I, 2.11).
Therefore, $X^{(H)}$ is a finite union of closed $G$-subspaces $G \times_K N^\nu$ where $N^\nu$ is the $(g_\nu^{-1}Hg_\nu)$-fixed point space of the $K$-module $N$. By assumption, $N^\nu$ is a $K$-ENR and using Corollary 6.6, we see that $N_i := \bigcup_{\nu \le i} N^\nu$ is one as well: For, $N_i$ consists of the closed $K$-subspaces $N_{i-1}$ and $N^i$, and their intersection is the $(g_i^{-1}Hg_i)$-fixed point space of $N_{i-1}$. Hence, $X^{(H)}$ is a $G$-ENR by Proposition 6.2.
We note that the locally closed $G$-subspace $M^{(H)} - \{0\} \subset M$ is a $G$-ENR as it is a local one. Let $S(M)$ denote the unit sphere in $M$. As a $G$-retract of some neighbourhood in $M^{(H)} - \{0\}$, $S(M)^{(H)}$ is a $G$-ENR. Of course, we assume that $G$ acts orthogonally on $M$. As a candidate for a $G$-ENR around the origin in $M^{(H)}$, we now take the part of $M^{(H)}$ in the closed unit ball. Since $G$ acts linearly, this is the cone over $S(M)^{(H)}$ which, in turn, is a $G$-ENR by Proposition 6.5.
The proof with $\underline{(H)}$ in place of $(H)$ proceeds by complete analogy. □

**6.8** The following application of the last two results will be used in Chapter III.

**Lemma.** *Let $p\colon E \to B$ be a $G$-ENR$_B$ with trivial $G$-action on its base. Then $p$ admits a finite filtration $\emptyset = p_0 \subset p_1 \subset \ldots \subset p_{r-1} \subset p_r = p$ by closed $G$-ENR$_B$s $p_i\colon E_i \to B$ in such a way that $E_i - E_{i-1}$ consists of all points in $E$ on orbits of a fixed type.*

PROOF. Since $p$ is a vertical $G$-ENR over a $G$-trivial base, there is only a finite number of orbit types on $E$, say $(H_1) \succ (H_2) \succ \ldots \succ (H_r)$ (see I, 1.2). We set $E_i := \bigcup_{\nu \le i} E^{(H_\nu)}$. Then $E_i - E_{i-1}$ is just $E_{(H_i)}$: For, if $(H_j)$ is any orbit type on $E_i - E_{i-1}$, then, on the one hand, there is some $\nu \le i$ such that $(H_j) \ge (H_\nu)$ and hence $j \le \nu$, while on the other hand, $j$ can not be strictly smaller than $i$. Whence $\nu = i$.
$E_i$ is closed in $E$ since all $E^{(H_\nu)}$ are, and finally, $p_i := p \,|\, E_i$ is a $G$-ENR$_B$: When $p$ is a vertical $G$-neighbourhood retract in the projection $B \times M \to B$, then $p_i$ is one in $B \times M_i \to B$ with $M_i := \bigcup_{\nu \le i} M^{(H_\nu)}$ because $G$ acts trivially on $B$. According to Proposition 6.7, each $M^{(H_\nu)}$ is a closed $G$-ENR in $M$. Hence, Corollary 6.6 implies via induction that $M_i$ is a $G$-ENR. □

**6.9** As a consequence, we obtain a sum formula for the Euler-Poincaré characteristic of a compact $G$-ENR which, in Chapter III, will be generalized to a formula for fixed point indices.
Let $\chi_c$ denote the Euler-Poincaré characteristic in singular cohomology with compact support. Since an ENR is locally contractible, we may calculate its characteristic in Alexander-Spanier cohomology ([Spanier], 6.9.5).

**Corollary.** *The Euler-Poincaré characteristic of a compact G-ENR E reads*

$$\chi(E) = \sum_{(H)} \chi_c(E_{(H)})$$

*where the sum is taken over all orbit types on E.*

PROOF. Consider the filtration 6.8 on $E$. Since each $E_i$ is a compact ENR, the Euler-Poincaré characteristics $\chi(E_i)$ and $\chi(E_i, E_{i-1})$ are defined, and from $\chi(E_i, E_{i-1}) = \chi(E_i) - \chi(E_{i-1})$ we get $\chi(E) = \Sigma\, \chi(E_i, E_{i-1})$. Calculating $\chi(E_i, E_{i-1})$ in Alexander-Spanier cohomology, we obtain $\chi_c(E_i - E_{i-1}) = \chi_c(E_{(H_i)})$ ([Spanier], 6.6.11). □

# CHAPTER III
# The Fixed Point Index of Equivariant Vertical Maps

By a *G-fixed point situation over B*, we mean a partially defined, vertical $G$-transformation $f$ on a $G$-ENR$_B$ whose fixed point set Fix($f$) lies properly over $B$. $G$ is supposed to be a *compact Lie group* and $B$ is provided as a *paracompact, compactly generated G-space*.
Two $G$-fixed point situations over $B$ are called equivalent if they can be connected through a $G$-fixed point situation over $B \times [0, 1]$. The corresponding equivalence classes form a commutative ring with unit, the *G-fixed point ring* $Fix_G(B)$. The *G-fixed point index* in a multiplicative $G$-cohomology theory $h_G$ is a unitary ring homomorphism $I_{h_G}$: $Fix_G(B) \to h_G^0(B)$ which enjoys the familiar properties as listed in [Dold 2] for the non-equivariant case.
In the **first four sections** of the present chapter, we define the $G$-fixed point index and derive its basic properties. Its typical equivariant character will be discussed in the **remaining sections.**

In the following, let $f$ be a $G$-fixed point situation over $B$. We start by constructing a *stable G-morphism* $i_G^f$: $B^+ \to B^+$, i.e. a morphism in the stable $G$-category, which has the usual index properties. In any *G-cohomology theory* $h_G$, $i_G^f$ induces an *endomorphism* $I_{h_G}^f$ *of* $h_G(B)$. Of course, we assume $B$ to be compact when $h_G$ has compact support. If $h_G$ is a *multiplicative* theory, $I_{h_G}^f$ is multiplication by the image $I_{h_G}(f)$ of the unit element in $h_G^0(B)$. We call $I_{h_G}(f) \in h_G^0(B)$ the *$h_G$-index of f*. It has the familiar properties. At the end of Section 2, we will define the *equivariant fixed point transfer* $T_{h_G}^f$ *of f* in order to study the behavior of the index under transfer maps. These are the contents of **Section 1** and **2.**

In **Section 3**, we investigate the *index $I_G(f)$ in stable G-cohomotopy theory* $\pi_G$. Any other $G$-cohomology theory $h_G$ is a $\pi_G$-module and because of the naturality of the index, $I_{h_G}^f$ is multiplication by $I_G(f)$. For illustration, we consider a $G$-fixed point situation over a pathwise connected base of finite category with trivial $G$-action. In this case, $I_{h_G}^f$ becomes an isomorphism when localized away from the index $I_G(f_{pt})$ of any fibre of $f$. In particular, the index of $f$ in a multiplicative theory $h_G$ is invertible if and only if $I_{h_G}(f_{pt})$ is so.
The $h_G$-index of the identity on a proper $G$-ENR$_B$ $p$ is called the *$h_G$-characteristic of p*, denoted by $\chi_{h_G}(p)$. We derive that the $\pi_G$-characteristic of a finite-sheeted $G$-covering is represented by the composite of its classifying map and the *equivariant group completion*

*map*. In the *Segal conjecture*, therefore, a finite $G$-set $X$ gets mapped to the $\pi$-characteristic of the covering projection $E(G) \times_G X \to B(G)$.
Finally, we show that the transfer of a compact, smooth $G$-manifold $X$ is induced by the Pontryagin-Thom construction for a $G$-embedding of $X$ into some $G$-module.

**Section 4** will be concerned with the *fixed point ring* $Fix_G(B)$. In a multiplicative $G$-cohomology theory, the index is a unitary ring homomorphism $I_{h_G}: Fix_G(B) \to h_G^0(B)$. Since stable $G$-cohomotopy is universal under multiplicative $G$-cohomology theories, the naturality of the index yields a commutative triangle

$$\begin{array}{ccc} & Fix_G(B) & \\ {\scriptstyle I_G}\swarrow & & \searrow{\scriptstyle I_{h_G}} \\ \pi_G^0(B) & \longrightarrow & h_G^0(B) \end{array}$$

We show that $I_G$ is an *isomorphism*. Therefore, precisely the stably spherical cohomology classes occur as fixed point indices in $h_G^0(B)$. Besides, we sketch the realization of stable G-cohomology classes of arbitrary degree $\alpha$ by $G$-fixed point situations of *degree* $\alpha$ whereby $\alpha$ is an element of the real representation ring of $G$.
Finally, we will interpret various maps between equivariant cohomology groups in terms of equivariant fixed point theory. So, let $\lambda: H \to G$ be a Lie homomorphism. If $B$ is an $H$-space on which the kernel of $\lambda$ acts freely, then $Fix_H(B)$ is isomorphic to $Fix_G(G \times_H B)$. For $\lambda$ the inclusion of a subgroup, in particular, this implies that any $G$-fixed point situation over $G/H$ is completely determined by its fibre over $H$ which clearly is an $H$-fixed point situation. Further, the $G$-index of a $G$-fixed point situation $f$ with free $G$-action corresponds to the Hopf index of $f/G$, the map induced on orbit spaces. The Segal map, for example, thus assigns to a finite $G$-set $X$ the $\pi_G$-characteristic of the projection $E(G) \times X \to E(G)$.
For $H \leq G$, the composite of the structure isomorphism $h_H^0(B) \to h_G^0(G \times_H B)$ with the $G$-transfer of the orbit projection $G \times_H B \to B/H$ is what we call *induction*. For $B$ a point, the induction becomes a homomorphism of coefficient groups

$$\mathrm{ind}_H^G: h_H^0(\mathrm{pt}) \to h_G^0(\mathrm{pt}).$$

In equivariant $K$-theory, in fact, $\mathrm{ind}_H^G$ is the classical induction homomorphism of complex representation rings as turns out in the last section.

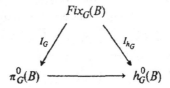

In **Section 5**, we specialize to $G$-fixed point situations over a base with trivial $G$-action. So, let $p: E \to B$ be a corresponding $G$-ENR$_B$. We first show that the $h_G$-index of a $G$-fixed point situation $f$ in $p$ decomposes into the sum

$$I_{h_G}(f) = \sum_{(H)} \left( I_{h_G}(f^{(H)}) - I_{h_G}(f^{\underline{(H)}}) \right)$$

taken over the orbit types $(G/H)$ around Fix($f$). The $(H)^{\text{th}}$ summand proves to be the index of a $G$-fixed point situation in $p_{(H)}$. For $B$ a point, it can be calculated exploiting that the projection $E_{(H)} \to E_{(H)}/G$ is a locally trivial fibre bundle with fibre $G/H$: It vanishes unless the $G$-automorphism group $W(H)$ of $G/H$ is finite. In that case, the Hopf index $I(f^H) - I(f^{\underline{H}})$ is divisible by the order of $W(H)$ yielding

$$I_{h_G}(f^{(H)}) - I_{h_G}(f^{\underline{(H)}}) = \left( I(f^H) - I(f^{\underline{H}}) \right) / \left| W(H) \right| \cdot \chi_{h_G}(G/H).$$

To summarize, the $h_G$-index of a $G$-fixed point situation $f$ over a point is an integral linear combination of the $h_G$-characteristics of the orbit types around its fixed point set, what will be referred to as the *sum formula for f*:

$$I_{h_G}(f) = \sum_{(H)} n_H(f) \cdot \chi_{h_G}(G/H).$$

In *ordinary singular cohomology*, it follows at once that the *Hopf index* $I(f)$ of $f$ equals $I(f^S)$ for every torus $S \leq G$. In particular, $I(f)$ vanishes if none of the isotropy groups on Fix($f$) is of maximal rank in $G$. Otherwise, $I(f)$ is divisible by the greatest common divisor of the *Euler-Poincaré characteristics* $\chi(G/H)$ of the orbits on Fix($f$). As an application, we will derive in the next section some Borsuk-Ulam theorems. In the case of a finite $p$-group $G$, finally, $I(f)$ is congruent to $I(f^G)$ modulo $p$.
We can calculate the coefficients in the sum formula as Lefschetz numbers if $f$ is a globally defined $G$-fixed point situation in a compact $G$-ENR $E$. For the identity on $E$, for instance, we get

$$n_H(f) = \chi_c(E_{(H)}) / \left| W(H) \right| = \chi_c(E_{(H)}/G)$$

where $\chi_c$ denotes the Euler-Poincaré characteristic in Čech or singular cohomology with compact support. The sum formula for the identity on $E$ corresponds exactly to the decomposition of the complex $G$-representation $\sum(-1)^i H^i(E;\mathbb{C})$ in [tom Dieck 2], 5.3.13. In fact, we will show in the last section that this representation of $G$ is nothing but the characteristic of $E$ in equivariant $K$-theory.

**Section 6** is dedicated to some *applications of the sum formula.* For $G$ finite, we deduce at once the following relation between Euler-Poincaré characteristics, originally due to T. tom Dieck ([tom Dieck 2], 5.3.12):

$$\sum_{g\in G} \chi(E^g) = \chi(E/G)|G|.$$

The superscript "$g$" stands for the subgroup generated by $g$. More generally, the sum of the Hopf indices $I(f^g)$ of a $G$-fixed point situation $f$ is the $|G|$-fold sum of the coefficients in our sum formula. For $G$ a compact Lie group again, we derive therefrom the *congruences*

$$\sum_{(K)} |\mathrm{gen}(K/H)|\ |N(H)/(N(H)\cap N(K))|\cdot I(f^K) \equiv 0 \mod |W(H)|$$

where the sum is taken over those conjugacy classes in $G$ which have a representative $K$ containing $H$ as a normal subgroup such that $K/H$ is cyclic. $H$ runs through the set of subgroups of $G$ with finite index in their normalizer. For the identity on a compact $G$-ENR, these congruences are known from [tom Dieck 2], 5.8.4. As an application, we list some *Borsuk-Ulam type consequences.*
Next, we reveal some *relationships between the Hopf indices of $f$ and $f/G$* - provided of course, the latter's fixed point set is compact. If $G$ is finite, we get

$$\sum_{g\in G} I(gf) = I(f/G)|G|.$$

Since the fixed point set of $f/G$ is the orbit space of $\bigcup_{g\in G}\mathrm{Fix}(gf)$, this is - generically - obvious when $G$ acts freely. And the general case can be reduced to this one by means of the sum formula.
For a $G$-fixed point situation in a $G$-ENR with a single orbit type $(G/H)$, we get

$$\sum_{\bar{g} \in G/H} I_{h_G}(\bar{g}f) = I(f/G) \cdot \chi_{h_G}(G/H)$$

provided $G$ is abelian. This relation holds for infinite $G$ as well and again, we can extend it to arbitrary $G$-ENRs using the sum formula.

A $G$-ENR with a single orbit type is a vertical $G$-ENR over its orbit space. If, in this case, $f/G$ is compactly fixed, there is a $G$-neighbourhood $U$ of $\mathrm{Fix}(f)$ without any other $f$-invariant orbit and we have

$$I_{h_G}(f) = I(f/G \mid U/G) \cdot \chi_{h_G}(G/H).$$

For the rest of the section, we let $G$ be a compact Lie group. On a compact $G$-ENR $E$, the multiplication by an element $g \in G$ has the Hopf index

$$I(g) = \chi(E^g).$$

For an element $g$ of finite order, this has been shown in [tom Dieck 2] and [Brown]. In that case, the statement follows readily from the relations at the beginning of the section which imply $\sum_{g \in G} I(g) = \sum_{g \in G} \chi(E^g)$ if $G$ is finite. Again, the general case reduces to the finite one via the sum formula. As an application, we derive three *folklore results on compact Lie groups*.

In **Section** 7, we study *local Hopf indices at regular fixed points*. If $f$ is a $G$-fixed point situation over a point such that the conjugacy classes of all of its isotropy groups around $\mathrm{Fix}(f)$ are finite, then, according to the equivariant transversality theorem presented in Chapter I, $f$ is equivalent to a smooth $G$-fixed point situation $\tilde{f}$ which has only regular fixed points. The local Hopf index of $\tilde{f}$ at some fixed point $x$ is the Hopf index of $T_x(\tilde{f})$, that is the sign of the determinant of $\mathrm{id} - T_x(\tilde{f})$. And the sum of these signs taken over all $x \in \mathrm{Fix}(\tilde{f})$ amounts to $I(\tilde{f}) = I(f)$. So, when investigating regular fixed points, we may confine ourselves to the case of a linear $G$-fixed point situation $f$ in some $G$-module. For elements $g$ and $h$ in $G$ commuting with each other, we then find

$$I(f^{gh}) / I(f) = I(f^g) \cdot I(f^h).$$

Thus, if $G$ is a finite abelian group, we get back the congruence $\sum_{g\in G} I(f^g) \equiv 0 \bmod |G|$ - even for global Hopf indices. If $G$ is a finite cyclic group, we deduce

$$I(f^H) \,/\, I(f) = \left(I(f^G) \,/\, I(f)\right)^{|G/H|}$$

for any $H \leq G$. Hence, $I(f^H)$ equals $I(f)$ in case $H$ does not contain the 2-Sylow subgroup of $G$. Otherwise, $I(f^H)$ equals $I(f^G)$ which holds as well if $G$ is merely abelian. Some examples will demonstrate that this result can not be improved.

**Section 8** is of theoretical nature again. We abbreviate the coefficient rings $Fix_G(\mathrm{pt})$ of equivariant fixed point theory by $F(G)$ and let $A(G) \leq F(G)$ denote the subring consisting of the identities on compact $G$-ENRs. The sum formula then says $A(G) = F(G)$. As an abelian group, in fact, $F(G)$ is free on the set $\Phi(G)$ of orbit types $(G/H)$ whose $G$-automorphism group $W(H)$ is finite.

Now, $f \mapsto \left(I(f^H)\right)$ defines a ring homomorphism $I^*$ from $F(G)$ to the mapping ring $\mathbb{Z}^{\Phi(G)}$. It proves to be injective. Therefore, $A(G) = F(G)$ is the classical *Burnside ring of* $G$. In particular, our ring isomorphism $I_G\colon F(G) \to \pi^0_G(\mathrm{pt})$ from Section 4 provides us with a *geometrical view of the tom Dieck isomorphism* $A(G) \cong \pi^0_G(\mathrm{pt})$ ([tom Dieck 2], 8.5.1.).

When $G$ is a finite group, the image of $F(G)$ in $\mathbb{Z}^{\Phi(G)} = \mathbb{Z}\cdot\Phi(G)$ is a subgroup of maximal rank. We can describe it by a *set of congruences* read off from the matrix $\left(|(G/H)^K|\right)$ of $I^*$. For this, we invert the matrix of $I^*$ - even if $G$ is not finite: We obtain a *three-dimensional matrix* $\mathrm{M} = \left(m^K_H(\mathcal{H})\right)$ with indices $H$, $K$, and $\mathcal{H}$, where $H$ and $K$ stand for representatives of the classes in $\Phi(G)$ and $\mathcal{H}$ for finite subsets of $\Phi(G)$. We will provide explicit formulae to calculate the entries in M.

The $\mathcal{H}^{\text{th}}$ layer $\mathrm{M}(\mathcal{H})$, an endomorphism of $\mathbb{Z}^{\Phi(G)}$, describes the image of the subgroup $F_{\mathcal{H}} \leq F(G)$ free on $\mathcal{H}$: A map $y \in \mathbb{Z}^{\Phi(G)}$ comes from $F_{\mathcal{H}}$ if and only if $\mathcal{H}$ is the support of the map $\mathrm{M}(\mathcal{H})\cdot y$ and $(\mathrm{M}(\mathcal{H})\cdot y)(H)$ is divisible by $|W(H)|$ for each $(H) \in \mathcal{H}$.

The subgroup $C_{\mathcal{H}} \leq \mathbb{Z}^{\Phi(G)}$, consisting of those maps $y$ for which $\mathrm{M}(\mathcal{H})\cdot y$ vanishes outside $\mathcal{H}$, is free on the maps $|W(H)|^{-1}\cdot I^*(G/H)$, $(H) \in \mathcal{H}$. In $C_{\mathcal{H}}$, therefore, the image of $F_{\mathcal{H}}$ is a subgroup of maximal rank and of index $\prod_{(H)\in\mathcal{H}} |W(H)|$. It is given by the congruences

$$(\mathrm{M}(\mathcal{H})\cdot y)(H) \equiv 0 \mod |W(H)|$$

for all $(H) \in \mathcal{H}$. In the union $C$ of all $C_{\mathcal{H}}$, the image of $F(G)$ is determined by the congruences

$$\sum_{(K)} |\text{gen}(K/H)|\, |N(H)/(N(H)\cap N(K))| \cdot y(K) \equiv 0 \mod |W(H)|$$

from Section 6. M reduces to a two-dimensional matrix in case $G$ is finite. And for $G$ a finite cyclic group $\langle g \rangle$, M yields the congruences

$$\sum_{P(k)} (-1)^{|P(k)|} \cdot y(\langle g^{k/P(k)} \rangle) \equiv 0 \mod k$$

where $k$ is any number dividing the order of $g$. The sum is taken over all subsets $P(k)$ of the set of prime factors in $k$ and $k/P(k)$ means $k$ divided by all $p \in P(k)$. Formally, these are the congruences between the Hopf indices of the *iterates of a fixed point situation g* found in [Dold 5].

In the **last section**, we will discuss the *fixed point index in equivariant K-theory*. We show that our induction homomorphism from Section 4 yields the induced representation as defined in [Segal 1] for compact Lie group actions. For the $K_G$-characteristic of a finite orbit $G/H$, in particular, we obtain the representation of $G$ induced by the trivial representation $\mathbb{C}$ of $H$, which is the permutation representation $\mathbb{C}(G/H)$.
Therefore, if $X$ is a compact $G$-ENR with a single orbit type $(G/H)$ which moreover is of finite length, the $K$-characteristic of the orbit projection $X \to X/G$ is represented by the vector bundle $X_H \times_{W(H)} \mathbb{C}(G/H) \to X/G$. And in case $G$ acts freely on $X$, the $K$-characteristic of the projection $X/H \to X/G$ is represented by the vector bundle $X \times_G \mathbb{C}(G/H) \to X/G$ for any subgroup $H$ of finite index in $G$.
Employing our sum formula, we then calculate the *$K_G$-index of a G-fixed point situation over a point:* As an element in the representation ring of $G$, it has the character function

$$I_{K_G}(f)\colon G \to \mathbb{Z}, \quad g \mapsto I(f^g).$$

This shows again that for $G$ finite, the sum of Hopf indices $\sum_{g\in G} I(f^g)$ is divisible by $|G|$. The $K_G$-characteristic of a compact $G$-ENR $E$, in particular, is the virtual representation

$$\chi_{K_G}(E) = \sum (-1)^i H^i(E;\mathbb{C}).$$

We end the chapter by describing the $K_G$-transfer of a compact, smooth $G$-manifold in terms of *Atiyah's topological index homomorphism* $t$-ind [Atiyah-Singer]. Its $K_G$-characteristic proves to be the image of its real Thom class under $t$-ind. Alternatively, this final result will be deduced from the *index formula of Atiyah and Singer.*

## 1. The Equivariant Stable Fixed Point Index

**1.1** Throughout, let $G$ be a *compact group* and $B$ a *paracompact* $G$-space whose topology is *compactly generated.*
Note that the total space of an $\mathrm{ENR}_B$, and hence every locally closed subspace therein, is then compactly generated - for the product of a compactly generated space with a locally compact space is compactly generated (see Appendix, A.7).

**1.2 Definition.** A *partial G-transformation* on a vertical $G$-space $p\colon E \to B$ is a vertical $G$-map $f\colon V \to E$, defined on some open $G$-subspace $V \subset E$. We say that $f$ is *compactly fixed* if its fixed point set $\mathrm{Fix}(f) = \{x \in V\colon f(x) = x\}$ lies properly over $B$. If, in addition, $p$ is a $G$-$\mathrm{ENR}_B$, we call $f$ a *G-fixed point situation in p.*

To any $G$-fixed point situation $f$ over $B$, we want to assign, in a canonical way, an endomorphism of the $G$-cohomology of the base $B$ with properties analogous to those of the index homomorphism in the non-equivariant setting. Toward this end, we first construct a stable $G$-transformation $i_G^f$ of $B^+$, the base with a discrete base point, and check its index properties in the *stable G-category* $\underline{\mathrm{Stab}}_G$. Again, we refer the reader to [Ulrich] for detailed proofs of the non-equivariant versions listed in [Dold 2], Section 2.

**1.3** The *objects in* $\underline{\mathrm{Stab}}_G$ are pairs $(X, \alpha)$ of pointed G-spaces $X$ with compactly generated topology and elements $\alpha$ of the real representation ring of $G$. The base point of $X$ has to be fixed under $G$. Instead of $(X, 0)$ we simply write $X$. If $i\colon A \to X$ is an inclusion of compactly generated $G$-spaces, the mapping cone of $i$, with the cone vertex as base point, yields an object $(X, A)$ in $\underline{\mathrm{Stab}}_G$ of degree 0. For a $G$-module $M$, for instance, $B \times (M, M - 0)$ is the Thom-space of the trivial bundle $B \times M \to B$, that is the *M-fold suspension* $\Sigma^M(B^+) = B^+ \wedge S^M$. The $M$-sphere $S^M$ is the one-point compactification of $M$, not the sphere in $M$ which we denote by $S(M)$.

The set of *morphisms in* Stab$_G$ from $(X,\alpha)$ to $(Y,\beta)$ is defined as the direct limit of the pointed $G$-homotopy groups $[X \wedge S^{\alpha\oplus L}, Y \wedge S^{\beta\oplus L}]_G^0$ taken over complex $G$-modules $L$ and complex $G$-monomorphisms. For $L$ sufficiently large, $\alpha \oplus L$ and $\beta \oplus L$ are genuine $G$-modules. Cofinality reasons make sure that the limit does not depend on the decomposition of $\alpha$ and $\beta$ into differences of real $G$-modules. It is built as follows:
In the category of complex $G$-modules and complex $G$-homomorphisms, $L$ has a complementary direct summand $M$ when embedded into some $N$. Suspending with $S^M$ then serves to set up a direct system:

$$[X \wedge S^{\alpha\oplus L}, Y \wedge S^{\beta\oplus L}]_G^0 \to [X \wedge S^{\alpha\oplus L\oplus M}, Y \wedge S^{\beta\oplus L\oplus M}]_G^0 = [X \wedge S^{\alpha\oplus N}, Y \wedge S^{\beta\oplus N}]_G^0$$

To show that this definition makes sense, i.e. that it does not depend on the particular choice of $M$, it suffices to consider the case $(X,\alpha) = S^0 = (Y,\beta)$. With $q: N \to L$ denoting the projection along $M$, the $M$-fold suspension of a $G$-map $\xi: (L, L-0) \to (L, L-0)$ is represented by the $G$-transformation $\zeta(x) := \xi(q(x)) + (x - q(x))$ of $(N, N-0)$. Now, $(L, L-0)$ and $(N, N-0)$ mean pairs of spaces rather than mapping cones. If $q': N \to L$ is the projection along some other complementary $G$-summand $M'$, then $x \mapsto q'(x) + (x - q(x))$ is a $G$-automorphism $\vartheta$ on $(N, N-0)$ with the property $q(\vartheta(x)) = q'(x)$, and the representative $\zeta'$ of the $M'$-fold suspension of $\xi$ maps some $x$ to $\xi(q'(x)) + (x - q'(x)) = \zeta(\vartheta(x)) + (x - \vartheta(x))$.
Now, for any $H \leq G$, the transformation of the $N^H$-sphere $(S^N)^H$ corresponding to $\vartheta^H$ has mapping degree 1, for the real determinant of a complex isomorphism is positive. Hence, the transformation of $S^N$ given by $\vartheta$ admits a pointed $G$-homotopy to the identity on $S^N$ ([tom Dieck 2], 8.4.1). The corresponding $G$-homotopy on $(N, N-0)$ from $\vartheta$ to the identity finally shows $[\zeta]_G^0 = [\zeta']_G^0$.

**1.4** In order to define the equivariant stable index morphism, we start with a $G$-fixed point situation $f: V \to E$ in a vertically trivial $G$-ENR$_B$ $p: E = B \times M \to B$ whose fibre is a complex $G$-module.
The fixed point set of $f$, lying properly over $B$, is closed in $E$ - even if $p$ is not vertically trivial - since $B$ is compactly generated (see Appendix, A.5). It is fibrewise uniformly bounded because $B$ is paracompact: There is a continuous function $\rho: B \to (0,\infty)$ such that Fix$(f)$ is contained in $E_\rho := \{(b,x) \in B \times M: \|x\| \leq \rho(b)\}$ (see Appendix, A.6). We may assume that $\rho$ is a $G$-function. For, since Fix$(f)$ is a $G$-space and since we may regard $M$ as a unitary $G$-module, $\rho'(b) := \int_G \rho(gb)$ can not be smaller than $\int_G \|gx\| = \|x\|$ as long as $(b,x)$ belongs to Fix$(f) \subset E_\rho$.

**Theorem and Definition:** *Let $f$ be a $G$-fixed point situation in a vertically trivial $G$-$\mathrm{ENR}_B$ $p: E = B \times M \to B$ with fibre $M$ a complex $G$-module, i.e. $f: V \to E$, $(b, x) \mapsto (b, \varphi(b, x))$. Then $i - f: V \to E$, $(b, x) \mapsto (b, x - \varphi(b, x))$ and appropriate inclusions define in* Stab$_G$ *an endomorphism $i_G^f$ of $B^+$ which we call the stable $G$-index morphism of $f$:*

$$
\begin{array}{ccccccc}
B \times (M, M-0) & & \xrightarrow{\qquad\qquad\qquad i_G^f \qquad\qquad\qquad} & & & & B \times (M, M-0) \\
\uparrow \sim & & & & & & \uparrow = \\
(E, E - E_\rho) & \xrightarrow{\subset} & (E, E - \mathrm{Fix}(f)) & \xleftarrow{\sim} & (V, V - \mathrm{Fix}(f)) & \xrightarrow{i-f} & B \times (M, M-0)
\end{array}
$$

Proof. The inclusion $(E, E - E_\rho) \to B \times (M, M - 0)$ of mapping cones, being a pointed $G$-homotopy equivalence, becomes an isomorphism in Stab$_G$. There, also the excision $(V, V - \mathrm{Fix}(f)) \to (E, E - \mathrm{Fix}(f))$ is an isomorphism: Like in the non-equivariant case (see [Dold-Puppe], 3.4), this follows via the extension theorem of Tietze-Gleason because all subspaces involved are open in $E$, and hence are compactly generated (1.1).
Finally, the definition does not depend on the choice of $\rho$ since we may replace $\rho$ by a greater function without causing any harm. □

In the following summary, *let $f: V \to E$ denote a $G$-fixed point situation in a vertically trivial $G$-$\mathrm{ENR}_B$ $p: E \to B$ with fibre a complex $G$-module $M$* - unless indicated otherwise.

**1.5 Stability** (STB). *Let $0$ denote the constant endormorphism of some $G$-module $N$. Then $f \times 0: V \times N \to E \times N$ is a $G$-fixed point situation over $B$ which has the same stable $G$-index morphism as $f$.*

Proof. At the moment, yet, we have to assume that $N$ is a complex $G$-module. On the one hand then, the $N$-fold suspension of sequence 1.4 represents in Stab$_G$ the same morphism as the sequence itself, that is $i_G^f$, whereas on the other hand, it just defines $i_G^{f \times 0}$. □

**1.6** The excision-property as well is obvious from the very definition:

**Localization** (LOC). *The stable $G$-index morphism of $f$ depends only on the germ of $f$ around* Fix$(f)$. To be precise: *The restriction of $f$ to any open $G$-neighbourhood of* Fix$(f)$ *in $V$ has the same stable $G$-index as $f$.* □

**1.7 Fixed Point Property** (FIX). In particular, *if $f$ does not have any fixed points,* $\mathrm{Fix}(f)=\emptyset$, *then $i_G^f$ is represented by the constant map* $+: B^+ \to B^+$. □

**1.8 Additivity** (ADD). *Suppose $f$ is the union of two partial $G$-transformations $f_1$ and $f_2$ of $p$ whose intersection is compactly fixed. Then $f_1$ and $f_2$ are compactly fixed.*
*In case $f_1$ and $f_2$ do not have any fixed points in common, the following diagram in* $\underline{\mathrm{Stab}}_G$ *is commutative where $\mu$ comes from the co-multiplication on* $\Sigma^M(B^+)$:

$$\begin{array}{ccc} B^+ & \xrightarrow{\quad i_G^f \quad} & B^+ \\ {\scriptstyle\mu}\big\downarrow & & \big\uparrow{\scriptstyle(\mathrm{id},\,\mathrm{id})} \\ B^+ \vee B^+ & \xrightarrow{\quad i_G^{f_1} \vee i_G^{f_2} \quad} & B^+ \vee B^+ \end{array}$$

PROOF. A straightforward argument shows that $f_1$ and $f_2$ are compactly fixed (see [Ulrich], p 32). Let $V_i$ be the domain and $F_i$ the fixed point set of $f_i$ for $i=1,2$. If $F_1 \cap F_2$ is empty, we may assume, after localization, that $V_1 \cap V_2$ is empty. The co-multiplication $\mu$ on the $M$-fold suspension $B^+ \wedge S^M$ constricts $B^+ \wedge S(M)$ to one point as $S(M)$ is the equator in $S^M = S(M \oplus \mathbb{R})$ (1.3). Therefore, $\mu$ is a $G$-map and it is obvious for geometric reasons (cf [Dold-Puppe], Proof 3.4) that the diagram below is commutative in $\underline{\mathrm{Stab}}_G$: $\mu'$ denotes the co-multiplication on the mapping cone $(E, E-E_\rho)$ which, in $\underline{\mathrm{Stab}}_G$, is isomorphic to $(E, E-0) = \Sigma^M(B^+)$. Arrows marked with a tilde are excision isomorphisms.

$$\begin{array}{ccccc} (E, E-E_\rho) & \xrightarrow{\ \subset\ } & (E, E-F) & \xleftarrow{\ \sim\ } & \\ & & \big\uparrow{\scriptstyle\sim} & & \\ {\scriptstyle\mu'}\big\downarrow & & (E-F_2\,, E-F) \vee (E-F_1\,, E-F) & & \\ & & \big\downarrow{\scriptstyle\sim} & & \\ (E, E-E_\rho) \vee (E, E-E_\rho) & \xrightarrow{\ \subset\ } & (E, E-F_1) \vee (E, E-F_2) & \xleftarrow{\ \sim\ } & \end{array}$$

$$\begin{array}{ccccc} \xleftarrow{\ \sim\ } & (V, V-F) & \xrightarrow{\quad i-f \quad} & \Sigma^M(B^+) \\ & {\scriptstyle=}\big\updownarrow & & \big\uparrow{\scriptstyle(\mathrm{id},\,\mathrm{id})} \\ \xleftarrow{\ \sim\ } & (V_1\,, V_1-F_1) \vee (V_2\,, V_2-F_2) & \xrightarrow{\ (i-f_1)\vee(i-f_2)\ } & \Sigma^M(B^+) \vee \Sigma^M(B^+) \end{array}$$

Since the top line represents $i_G^f$ and the bottom line $i_G^{f_1} \vee i_G^{f_2}$, the result follows. □

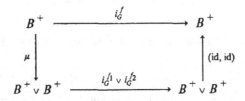

**1.9 Addendum.** *In general, i.e. when we allow $f_1$ and $f_2$ to have common fixed points, we rather get the following commutative diagram in* $\underline{\mathrm{Stab}}_G$:

$$\begin{array}{ccccc} & B^+\vee B^+ & \xrightarrow{\;i_G^{f}\,\vee\, i_G^{f_1\cap f_2}\;} & B^+\vee B^+ & \\ {\scriptstyle\mu}\nearrow & & & & \searrow{\scriptstyle(\mathrm{id},\,\mathrm{id})} \\ B^+ & & & & B^+ \\ {\scriptstyle\mu}\searrow & & & & \nearrow{\scriptstyle(\mathrm{id},\,\mathrm{id})} \\ & B^+\vee B^+ & \xrightarrow{\;i_G^{f_1}\,\vee\, i_G^{f_2}\;} & B^+\vee B^+ & \end{array}$$

Observe that for $\mathrm{Fix}(f_1\cap f_2)=\emptyset$, the upper half reduces to $i_G^f$ because of FIX.

PROOF. We set $f' := f\,|\,V-(F_1\cap F_2)$ and $f_i' := f'\,|\,V_i$, and add two lines to the above diagram. The statement then follows by applying ADD in the four cases $f=(f_1\cap f_2)\cup f'$, $f'=f_1'\cup f_2'$, and $f_i=(f_1\cap f_2)\cup f_i'$ for $i=1,2$:

$$\begin{array}{ccc} B^+\vee B^+ & \xrightarrow{\;i_G^{f}\,\vee\, i_G^{f_1\cap f_2}\;} & B^+\vee B^+ \\ {\scriptstyle\mu\vee\mathrm{id}}\big\downarrow & & \big\uparrow{\scriptstyle(\mathrm{id},\,\mathrm{id})\vee\mathrm{id}} \\ B^+\vee B^+\vee B^+ & \xrightarrow{\;i_G^{f_1\cap f_2}\,\vee\, i_G^{f'}\,\vee\, i_G^{f_1\cap f_2}\;} & B^+\vee B^+\vee B^+ \\ {\scriptstyle\mathrm{id}\vee\mu\vee\mathrm{id}}\big\downarrow & & \big\uparrow{\scriptstyle\mathrm{id}\vee(\mathrm{id},\,\mathrm{id})\vee\mathrm{id}} \\ B^+\vee B^+\vee B^+\vee B^+ & \xrightarrow{\;i_G^{f_1\cap f_2}\,\vee\, i_G^{f_1'}\,\vee\, i_G^{f_2'}\,\vee\, i_G^{f_1\cap f_2}\;} & B^+\vee B^+\vee B^+\vee B^+ \\ {\scriptstyle\mu\vee\mu}\big\uparrow & & \big\downarrow{\scriptstyle(\mathrm{id},\,\mathrm{id})\vee(\mathrm{id},\,\mathrm{id})} \\ B^+\vee B^+ & \xrightarrow{\;i_G^{f_1}\,\vee\, i_G^{f_2}\;} & B^+\vee B^+ \end{array}$$

□

**1.10 Pull-back (PUB).** *The pull-back $b^*(f)$ of $f$ under an equivariant change of base $b\colon B'\to B$ is a G-fixed point situation over $B'$ and we have*

$$i_G^f\circ b^+ = b^+\circ i_G^{b^*(f)}.$$

This naturality in $B$ is proved like in the non-equivariant case. □

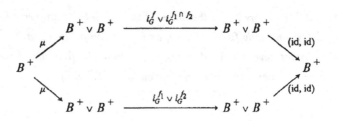

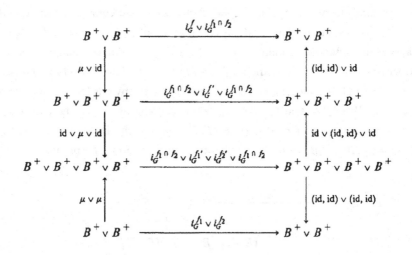

**1.11** The following is the fundamental property of the fixed point index:

**Homotopy Invariance** (HTP). *Let $f$ be a $G$-fixed point situation over $B \times [0, 1]$. For every $t \in [0, 1]$, the part $f_t$ of $f$ over $B \times \{t\}$ is a $G$-fixed point situation over $B$ whose stable $G$-index morphism does not depend on $t$.*

PROOF. $f_t$ is the pull back of $f$ under the equivariant change of base $b_t : B \approx B \times \{t\} \to B \times [0, 1]$. The assertion follows from $i_G^f \circ b_t^+ = b_t^+ \circ i_G^{f_t}$ (PUB) observing that the projection $B \times [0, 1] \to B$ is a $G$-homotopy inverse for $b_t$. □

**1.12 Multiplicativity** (MUL). *Let $f$ and $f'$ be $G$-fixed point situations over $B$ in $p$ and $p'$ respectively. Then their fibrewise product $f \times_B f'$ is a $G$-fixed point situation in $p \times_B p'$ over $B$ whose stable $G$-index morphism reads*

$$i_G^f \circ i_G^{f'} = i_G^{f \times_B f'} = i_G^{f'} \circ i_G^f.$$

PROOF. Like in the non-equivariant setting, the endomorphism $i_G^{f \times_B f'}$ of $\Sigma^{M \oplus M'}(B^+) = \Sigma^M(B^+) \wedge_B \Sigma^{M'}(B^+)$ in $\underline{\mathrm{Stab}}_G$ comes out as the diagonal in the following commutative diagram:

$$\begin{array}{ccc} \Sigma^M(B^+) \wedge_B \Sigma^{M'}(B^+) & \xrightarrow{\;i_G^f \wedge_B \mathrm{id}\;} & \Sigma^M(B^+) \wedge_B \Sigma^{M'}(B^+) \\ {\scriptstyle \mathrm{id} \wedge_B i_G^{f'}} \Big\downarrow & & \Big\downarrow {\scriptstyle \mathrm{id} \wedge_B i_G^{f'}} \\ \Sigma^M(B^+) \wedge_B \Sigma^{M'}(B^+) & \xrightarrow{\;i_G^f \wedge_B \mathrm{id}\;} & \Sigma^M(B^+) \wedge_B \Sigma^{M'}(B^+) \end{array}$$

Since $M$ and $M'$ are complex $G$-modules, the horizontal arrows both define $i_G^f$ and the vertical ones $i_G^{f'}$. □

**1.13 Commutativity** (COM). *Let $f: U \to E'$ and $f': U' \to E$ be vertical $G$-maps defined on open $G$-subspaces of $G$-ENR$_B$s $p: E \to B$ and $p': E' \to B$ respectively.*
*If one of the two partial $G$-transformations $f'f$ of $p$ and $ff'$ of $p'$ is compactly fixed, then so is the other and their stable $G$-index morphisms coincide:*

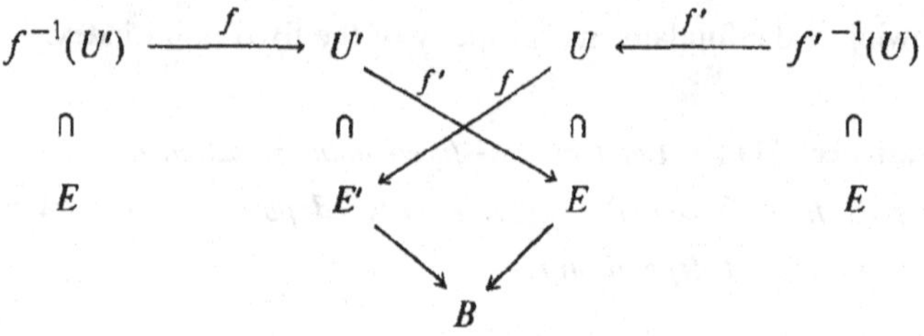


$$i_G^{f'f} = i_G^{ff'}$$

PROOF. Obviously, $f\colon \mathrm{Fix}(f'f) \to \mathrm{Fix}(ff')$ is a homeomorphism. For $i_G^{f'f} = i_G^{ff'}$, we show that $f'f$ and $ff'$ both have the same stable $G$-index morphism as the $G$-fixed point situation $\xi\colon U \times_B U' \to E \times_B E'$, $(x, x') \mapsto (f'(x'), f(x))$. This in turn is an application of HTP: On $f^{-1}(U') \times_B U' \subset U \times_B U'$, we connect $\xi$ to the localization of the vertical $G$-fixed point situation $\zeta\colon f^{-1}(U') \times_B E' \to E \times_B E'$, $(x, x') \mapsto (f'f(x), f(x))$ via a linear homotopy in the euclidean component $M$ of $E = B \times M$. And a linear homotopy in the euclidean component $M'$ of $E'$ leads onward from $\zeta$ to $f'f \times 0\colon f^{-1}(U') \times M' \to E \times M'$.
Since $G$ acts linearly on $M$ and $M'$, both homotopies are $G$-maps. They are compactly fixed as maps over $B \times [0, 1]$ because $f'f$ is so over $B$. HTP and LOC then imply $i_G^{\xi} = i_G^{\zeta}$ and HTP and STB yield $i_G^{\zeta} = i_G^{f'f}$.
Analogously, connecting $\xi$ first to the $G$-fixed point situation $(x, x') \mapsto (f'(x'), ff'(x'))$ and then to $0 \times ff'$, we get $i_G^{\xi} = i_G^{ff'}$. □

We append three immediate consequences of COM.

**1.14 Identity** (IDT). *For a G-cross section s in a G-*$\mathrm{ENR}_B$ *p,* $i_G^{sp}$ *is the class of the identity on* $B^+$.

For, the latter obviously represents the stable $G$-index morphism of the identity on $B$. □

**1.15 Topological Invariance** (TOP). *Let* $\eta\colon p \to p'$ *be a vertical G-homeomorphism between G-*$\mathrm{ENR}_B$*s. Then a G-fixed point situation f in p has the same stable G-index morphism as the G-fixed point situation* $\eta f \eta^{-1}$ *in* $p'$. □

**1.16 Shrinking Invariance** (SHR). *A G-fixed point situation* $f\colon V \to E$ *over B, which ranges in a G-*$\mathrm{ENR}_B$ $E' \subset E$, *has the same stable G-index morphism as its shrinking* $f_{\mathrm{shr}}\colon V \cap E' \to E'$. □

**1.17** So far, equivariant index theory is confined to $G$-fixed point situations in vertically trivial $G$-ENR$_B$s with fibre a complex $G$-module. Since any $G$-module embeds as a $G$-retract into some complex $G$-module - $N$ for instance as $N \times \{0\}$ into $N \oplus N$ - we can employ COM to extend the theory to the general case.

**Proposition and Definition.** *Let $p: E \to B$ be a $G$-ENR$_B$ over a paracompact, compactly generated base with $G$ a compact group.*
*Every partial $G$-transformation $f$ of $p$ decomposes into $G$-mappings over $B$, $f = \xi\zeta: V \to W \to E$ such that $W$ is an open $G$-subspace in a vertically trivial $G$-ENR$_B$ proj: $B \times M \to B$ with fibre $M$ a complex $G$-module.*
*If $f$ is compactly fixed, then the reverse composition $\zeta\xi$ is a $G$-fixed point situation in* proj *whose stable $G$-index morphism is defined by sequence 1.4. It depends only on $f$ not on the decomposition of $f$ chosen.*
So, we may call it the *stable $G$-index morphism of $f$*, denoted by $i_G^f: B^+ \to B^+$. □

**1.18** In particular, $i_G^f$ can be calculated as follows:

**Corollary.** *Let $f$ be a $G$-fixed point situation in a $G$-ENR$_B$ $p$. If $i$ and $r$ are the data of a vertical $G$-neighbouhood retraction onto $p$, i.e. $r\,i = \mathrm{id}_p$, in some vertically trivial $G$-ENR$_B$ (with a complex $G$-module as fibre), then $i_G^f$ is the stable $G$-index morphism of $i\,f\,r$.* □

**1.19** As in the non-equivariant case, we get:

**Proposition.** *The properties 1.5 to 1.16 continue to hold for $G$-fixed point situations in arbitrary $G$-ENR$_B$s.* □

## 2. The Equivariant Cohomological Fixed Point Index

**2.1** Throughout the section, let $G$ denote a *compact Lie group*. As before, the base $B$ is supposed to be paracompact and compactly generated.

**2.2** By a *reduced $G$-cohomology theory*, we understand, following [Kosniowski], a family of contravariant functors $\tilde{h}_G^*$ from a suitable category of pointed $G$-spaces - as for example compact or cellular ones - to the category of abelian groups. The functors are graded by

elements $\alpha$ of the real representation ring of $G$. They satisfy the homotopy and the exactness axiom, and for an irreducible $G$-module $M$, they come with natural isomorphisms $\sigma^{\alpha,M}\colon \tilde{h}_G^{\alpha} \to \tilde{h}_G^{\alpha+M} \circ \Sigma^M$ which are to commute in the graded sense.
For an arbitrary $G$-module $M$, we may get *two* suspension isomorphisms, differing by sign, depending on the decomposition of $M$ into irreducible representations. *Let us now fix, once and for all, one suspension isomorphism $\sigma^{\alpha,M}$ for each G-module $M$.*

We speak of an *equivariant cohomology theory* if, in addition, $\tilde{h}_G^{*}$ is a contravariant functor in the variable $G$: For this, we allow $G$ to vary in some category $\underline{G}$ of compact Lie groups with the property that $\underline{G}$ contains all inclusions of subgroups into each of its objects. A Lie homomorphism $\lambda\colon H \to G$ in $\underline{G}$ induces a natural transformation $\theta^{\lambda}\colon \tilde{h}_G^{*} \to \tilde{h}_H^{\lambda^*(\alpha)}$ of $G$-$H$-cohomology theories. We demand that the composite homomorphism

$$\tilde{h}_G^{\alpha}(G/H) \xrightarrow{\theta^{i}} \tilde{h}_H^{(\alpha\mid H)}(G/H) \xrightarrow{(i/H)^*} \tilde{h}_H^{(\alpha\mid H)}(\mathrm{pt})$$

be an isomorphism for every subgroup $i\colon H \leq G$. Then, for any $H$-space $B$ on which the kernel of $\lambda\colon H \to G$ acts freely, the composite

$$\tilde{h}_G^{\alpha}(G \times_H B) \xrightarrow{\theta^{\lambda}} \tilde{h}_H^{\lambda^*(\alpha)}(G \times_H B) \xrightarrow{(\lambda \times_H \mathrm{id}_B)^*} \tilde{h}_H^{\lambda^*(\alpha)}(B)$$

is an isomorphism, too ([Kosniowski], 2.14).

The corresponding unreduced theory on pairs of $G$-spaces will be denoted by $h_G$. Thus, $h_G(B)$ equals $\tilde{h}_G(B^+)$, and for well-based $G$-spaces, we have $\tilde{h}_G(X) \cong h_G(X, \mathrm{pt})$.

**2.3 Definition.** Let $f$ be a $G$-fixed point situation over $B$. By its *index homomorphism* in a $G$-cohomology theory $h_G$, we mean the *endomorphism $I^f_{h_G}$ of $h_G(B)$* induced by the stable G-index morphism $i^f_G\colon B^+ \to B^+$.

**2.4 Remark.** If $h_G$ is defined on compact or cellular $G$-spaces only, then $B$ has to be a member of the respective category.
Further, definition 2.3 does not depend on the choice of the suspension isomorphisms $\sigma^{\alpha,M}$ since in the sequence $h_G(i^f_G)$ induced by 1.4, the isomorphism used to suspend at the beginning is likewise used to desuspend at the end.

**2.5** The properties 1.5 to 1.16 of the stable $G$-index morphism imply at once the properties of the cohomological index homomorphism well-known for the non-equivariant setting. As to ADD, recall that the co-multiplication $\mu$ on $\Sigma^M(B^+)$ induces the addition in cohomology, i.e. the composite

$$h_G^\alpha(B) \oplus h_G^\alpha(B) \cong \widetilde{h}_G^{\alpha+M}(\Sigma^M B^+ \vee \Sigma^M B^+) \xrightarrow{\mu^*} \widetilde{h}_G^{\alpha+M}(\Sigma^M B^+) \cong h_G^\alpha(B)$$

maps $(b, b')$ to $b + b'$, and that the composite

$$h_G^\alpha(B) \cong \widetilde{h}_G^{\alpha+M}(\Sigma^M B^+) \xrightarrow{(\mathrm{id},\, \mathrm{id})^*} \widetilde{h}_G^{\alpha+M}(\Sigma^M B^+ \vee \Sigma^M B^+) \cong h_G^\alpha(B) \oplus h_G^\alpha(B)$$

is the diagonal mapping $b \mapsto (b, b)$.

**Theorem.** *The index homomorphism in a $G$-cohomology theory $h_G$ has the properties*

$$\text{HTP} \quad I_{h_G}^{f_1} = I_{h_G}^{f_0}$$

| | | | |
|---|---|---|---|
| STB | $I_{h_G}^{f\times 0} = I_{h_G}^{f}$ | MUL | $I_{h_G}^{f} \circ I_{h_G}^{f'} = I_{h_G}^{f\times_B f'} = I_{h_G}^{f'} \circ I_{h_G}^{f}$ |
| LOC | $I_{h_G}^{f\mid V'} = I_{h_G}^{f}$ | COM | $I_{h_G}^{f'f} = I_{h_G}^{ff'}$ |
| FIX | $I_{h_G}^{\emptyset} = 0$ | IDT | $I_{h_G}^{1_P} = \mathrm{id}_{h_G(B)}$ |
| ADD | $I_{h_G}^{f_1\cup f_2} = I_{h_G}^{f_1} + I_{h_G}^{f_2} - I_{h_G}^{f_1\cap f_2}$ | TOP | $I_{h_G}^{\eta f \eta^{-1}} = I_{h_G}^{f}$ |
| PUB | $I_{h_G}^{b^*(f)} \circ h_G(b) = h_G(b) \circ I_{h_G}^{f}$ | SHR | $I_{h_G}^{f_{\mathrm{shr}}} = I_{h_G}^{f}$ |

Again, the base is provided as a compact or a cellular $G$-space in case $h_G$ is confined to the respective category. □

**2.6 Naturality** (NAT). *Let $\theta^\lambda\colon h_G \to k_H^{\lambda^*}$ be a natural transformation of $G$-$H$-cohomology theories with $\lambda\colon H \to G$ a Lie homomorphism.*
*By virtue of $\lambda$, any $G$-fixed point situation $f$ over $B$ can be regarded as an $H$-fixed point situation over $B$, and the equivariant index homomorphism of $f$ commutes with $\theta$:*

$$I_{h_G}^{f} \circ \theta^\lambda = \theta^\lambda \circ I_{k_H}^{\lambda^*(f)}.$$

PROOF. This is obvious from the definition of the index homomorphism since natural transformations commute both with induced maps and suspension isomorphisms. □

**2.7** Next, let $h_G$ be a *multiplicative G-cohomology theory*. Then the cohomology of any vertical $G$-space $E$ over $B$ is a module over the ring $h_G(B)$ and a $G$-map over $B$ induces a homomorphism of $h_G(B)$-modules.

As known from the non-equivariant setting, the suspension isomorphism $\sigma^{\alpha,M}: \tilde{h}_G^{\alpha}(X) \to \tilde{h}_G^{\alpha+M}(\Sigma^M X) \cong h_G^{\alpha+M}(X \times (M, M-0))$ is multiplication by the spherical class $s^M = \sigma^{0,M}(1) \in \tilde{h}_G^M(S^M) \cong h_G^M(M, M-0)$ of degree $M$. Thus, when $X$ is a vertical $G$-space over $B$, $\sigma^{\alpha,M}$ is a graded homomorphism of $h_G(B)$-modules.

Since the defining sequence 1.4 for the stable $G$-index morphism is composed of vertical $G$-maps, the index homomorphism $I_{h_G}^f: h_G(B) \to h_G(B)$ becomes a homomorphism of $h_G(B)$-modules. Denoting the image of the unit element by $I_{h_G}(f)$, we have shown:

**The Index Element** (IND). *In a multiplicative G-cohomology theory $h_G$, the index homomorphism of a G-fixed point situation f over B is multiplication by the image $I_{h_G}(f)$ of the unit element $1_B \in h_G(B)$.*

We call $I_{h_G}(f) \in h_G^0(B)$ the *index of f* in the cohomology theory $h_G$. By the *$h_G$-characteristic of a proper* $G$-ENR$_B$ $p$, denoted $\chi_{h_G}(p)$, we mean the $h_G$-index of the identity on $p$. □

**2.8** We end with the behaviour of the index element under the fixed point transfer. To define the equivariant transfer, we start again with a $G$-fixed point situation $f$ in a vertically trivial $G$-ENR$_B$ $p$. In sequence 1.4, the final map $i-f$ decomposes as

$$i - f = (p \times \mathrm{id}_M) \circ (\mathrm{id}_V, i - f)_B :$$

$$(V, V - \mathrm{Fix}(f)) \longrightarrow V \times_B (B \times (M, M-0)) \approx V \times (M, M-0) \longrightarrow B \times (M, M-0).$$

Following [Dold 3], we propose

**Definition and Proposition.** Let $f: V \to E$ be a $G$-fixed point situation in a vertically trivial $G$-ENR$_B$ whose fibre is a complex $G$-module $M$. The part of sequence 1.4 up to $V \times (M, M-0)$ inclusively defines a morphism $t_G^{f,V}: B^+ \to V^+$ in $\underline{\mathrm{Stab}}_G$ which we call the *stable G-transfer of f*.

*It enjoys properties which are completely analogous to those of the stable G-index morphism. In particular, via the commutativity $f'^{+}\circ t_G^{f'f} = t_G^{ff'}$, we can extends its definition to G-fixed point situations in arbitrary G-ENR$_B$s.*

The stable $G$-transfer of (the identity on) a proper $G$-ENR$_B$ $p: E \to B$ is denoted by $t_G^p: B^+ \to E^+$. □

**2.9 Definition.** Let $f: V \to E$ be a $G$-fixed point situation over $B$ and let $h_G$ be any $G$-cohomology theory. *In case* the stable $G$-transfer $t_G^{f,V}: B^+ \to V^+$ of $f$ induces a homomorphism in $h_G$, we denote that by $T_{h_G}^{f,V}: h_G(V) \to h_G(B)$ and call it the *$h_G$-transfer of $f$*. *For example,* in a $G$-cohomology theory confined to compact $G$-spaces, the transfer $T_{h_G}^p: h_G(E) \to h_G(B)$ of a proper $G$-ENR$_B$ $p: E \to B$ is declared *provided $B$ is compact.*

**2.10** Remark. The stable $G$-index morphism of a $G$-fixed point situation $f$ in $p$ reads $(p|V)^+ \circ t_G^{f,V}$. In particular, $T_{h_G}^{f,V} \circ (p|V)^*$ *is the $h_G$-index homomorphism of $f$*. In a multiplicative $G$-cohomology theory, therefore, *the transfer of $f$ maps the unit element of $h_G(V)$ to the index $I_{h_G}(f)$ of $f$.*

**2.11** The following naturality of the index under transfer maps is a simple special case of the multiplicative behavior of the transfer as pointed out in [Dold 3], 8.7:

**Proposition.** *Let $p: E \to C$ and $q: C \to B$ be vertical $G$-ENRs and suppose $q$ is a proper map. Then any $G$-fixed point situation $f$ in $p$ can be regarded as a $G$-fixed point situation $q_*(f)$ in the $G$-ENR$_B$ $q_*(p) = qp$. In a $G$-cohomology theory $h_G$ in which the transfer of $q$ is defined, the index homomorpism of $q_*(f)$ is the composite*

$$h_G(B) \xrightarrow{q^*} h_G(C) \xrightarrow{I_{h_G}^f} h_G(C) \xrightarrow{T_{h_G}^q} h_G(B).$$

*In particular, the $h_G$-transfer of $q$ maps the index of $f$ to the index of $q_*(f)$ if $h_G$ is a multiplicative theory.*

Proof. $q_*(p)$ is a $G$-ENR$_B$ (II, 1.3), and $q_*(f)$ is compactly fixed if $f$ is, because $C$ lies properly over $B$.

By virtue of COM, we may assume that $p$ is a projection $E = C \times N \to C$. Write $F$ for $\mathrm{Fix}(f) = \mathrm{Fix}(q_*(f))$ and let $r: O \to C$ be a vertical $G$-neighbourhood retraction in $B \times M$. We may take complex $G$-modules for $M$ and $N$.

Now, choose a representative for $t_G^q$ (2.8), form its $N$-fold suspension, and interchange $M$ and $N$: The result is the left hand side in the following diagram:

$$
\begin{array}{ccccc}
 & & B\times(M\times N, M\times N-0) & & \\
 & \nearrow\sim & & \nwarrow\sim & \\
(B\times M\times N, B\times M\times N-(B\times M)_\sigma\times 0) & & \xleftarrow{\ \sim\ } & & (B\times M\times N, B\times M\times N-(B\times M\times N) \\
\downarrow\subset & & & & \downarrow\subset \\
(B\times M\times N, B\times M\times N-C\times 0) & & & & (B\times M\times N, B\times M\times N-F) \\
\uparrow\sim & & & & \uparrow\sim \\
(O\times N, O\times N-C\times 0) & & & & (O\times N, O\times N-F) \\
\downarrow & & & & \downarrow \\
C\times(N\times M, N\times M-0) & & & & (C\times N\times M, C\times N\times M-F\times 0) \\
 & \nwarrow\sim & & \nearrow\subset & \\
 & & \left(C\times N\times M, C\times N\times M-(C\times N)_\rho\times 0\right) & &
\end{array}
$$

Inclusions marked with a tilde are excision isomorphisms. Both of the unnamed arrows have, as maps ranging in the fibre product $(C\times N)\times_B(B\times M)$, the components $r\times\mathrm{id}_N$ and $(i_q-r)\circ\mathrm{proj}_O$ with $i_q$ the inclusion $O\to B\times M$. $\sigma$ encloses the space $C$ which lies properly over $B$, and $\rho$ does the same with $F$ over $C$.

In order to make $\tau$ a tubular neighbourhood of $F$ over $B$, we set $\tau(b):=\sqrt{(\sigma(b))^2+(\max\{\rho(x): x\in q^{-1}(b)\})^2}$. $\tau$ is well-defined and upper semi-continuous since the fibres of $q$ are compact. And $\tau$ is lower semi-continuous because $q$ is an open map (II, 3.3): For, if $\rho(x)$ is the maximum of $\tau$ on the fibre $q^{-1}(q(x))$ and if $\varepsilon>0$ is given, we choose a neighbourhood $U\subset C$ of $x$ on which $\rho$ stays above $\rho(x)-\varepsilon$. Then $q(U)$ is a neighbourhood of $q(x)$ and in every fibre over $q(U)$, the maximum of $\rho$ is greater than $\rho(x)-\varepsilon$.

Obviously, our diagram commutes. Up to excision, the right hand map $(r\times\mathrm{id}_N, (i_q-r)\circ\mathrm{proj}_O)_B$ coincides with its restriction to the mapping cone $\left((r\times\mathrm{id}_N)^{-1}(V), (r\times\mathrm{id}_N)^{-1}(V)-F\right)$. The latter gets mapped to $(V\times M, V\times M-F\times 0)$, and the open square

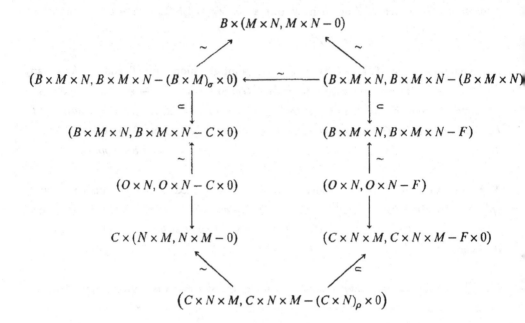

$$\begin{array}{ccc} \left((r\times \mathrm{id}_N)^{-1}(V),\, (r\times \mathrm{id}_N)^{-1}(V)-F\right) & \xrightarrow{(i\times \mathrm{id}_N)-((i\times \mathrm{id}_N)f(r\times \mathrm{id}_N))} & B\times (M\times N, M\times N-0) \\ \downarrow & & \\ (V\times M, V\times M - F\times 0) & \xrightarrow{(i_p - f)\times \mathrm{id}_M} & C\times (N\times M, N\times M-0) \end{array}$$

with the inclusions $i\colon C\to B\times M$, and $i_p\colon V\to C\times N$ becomes commutative when we insert the product of $q\colon C\to B$ with the interchanging $\vartheta$ of $N$ and $M$. Combining the two diagrams, we obtain a commutative square of the form

$$\begin{array}{ccc} B\times (M\times N, M\times N-0) & \longrightarrow & B\times (M\times N, M\times N-0) \\ \downarrow & & \uparrow \\ C\times (N\times M, N\times M-0) & \longrightarrow & C\times (N\times M, N\times M-0) \end{array}$$

The lower edge represents the stable $G$-index morphism of $f$ and the upper one that of $q_*(f)$, and apart from $\vartheta$, the right edge is $q$ and the left one represents the stable $G$-transfer of $q$. The assertion is proved since in cohomology, the interchange $\vartheta$ results in a sign. □

## 3. The Index in Equivariant Stable Cohomotopy Theory

**3.1** *Equivariant stable cohomotopy groups* constitute an equivariant multiplicative cohomology theory $\pi_G$. On any pointed $G$-space $X$, $\tilde{\pi}_G^{\alpha}(X)$ is defined as the morphism set $\underline{\mathrm{Stab}}_G((X,0),(S^0,\alpha))$, which is the direct limit of the pointed $G$-homotopy groups $[X\wedge S^L, S^{\alpha\oplus L}]_G^0$ taken over complex $G$-modules $L$ and complex $G$-monomorphisms (1.3). As before, $\alpha$ is a virtual representation of $G$ over $\mathbb{R}$.

**3.2** *Any other $G$-cohomology theory $h_G$ is a graded $\pi_G$-module:* An element $b\in \tilde{\pi}_G^{\beta}(B^+)=\pi_G^{\beta}(B)$, represented by a pointed $G$-map $b\colon B^+\wedge S^N\to S^M$ with $M-N=\beta$, acts on $h_G(B)$ via the homomorphism $(\mathrm{proj}_B, b)^*$:

$$(b\cdot -): h_G^{\alpha}(B)\cong \tilde{h}_G^{\alpha+M}(B^+\wedge S^M)\to \tilde{h}_G^{\alpha+M}(B^+\wedge S^N)\cong h_G^{\alpha+\beta}(B).$$

When $b$ represents the $\pi_G$-index of some $G$-fixed point situation $f$ over $B$, then $(\mathrm{proj}_B, b)$ is a representative for the stable $G$-index morphism of $f$. Hence

**Corollary.** *Let $h_G$ be any $G$-cohomology theory. The index homomorphism of a $G$-fixed point situation $f$ over $B$ is the multiplication by $I_G(f)$, the $\pi_G$-index of $f$:*

$$I^f_{h_G} = (I_G(f)\cdot -) : h_G(B) \to h_G(B).$$

*In particular, $I^f_{h_G}$ is a $\pi_G$-endomorphism.* □

**3.3 Theorem.** *Among all $G$-cohomology theories $h_G$, $\pi_G$ has the following universal property: For every $x \in h^0_G(\mathrm{pt})$, there exists a unique natural transformation $\hat{x}: \pi_G \to h_G$ of $G$-cohomology theories which maps the unit element of $\pi_G$ to $x$. In fact, for some $b: B^+ \wedge S^N \to S^M$ in $\pi^{M-N}_G(B)$, $\hat{x}(b) \in h^{M-N}_G(B)$ is given by $\sigma^{M,-N} \circ b^* \circ \sigma^{0,M}(x) = b\cdot((B \to \mathrm{pt})^*(x))$.*
*If $h_G$ is a multiplicative theory, then the natural transformation corresponding to the unit element is multiplicative.* We denote it by $u: \pi_G \to h_G$ and call the cohomology classes in its image *stably spherical.* □

**3.4** COMMENTS. In other words, *$\pi_G$ is an initial object in the category of multiplicative $G$-cohomology theories $h_G$.*
Occasionally, $u: \pi_G(B) \to h_G(B)$ is referred to as the *degree map*, for

$$u(b) = b\cdot 1_B = \sigma^{M,-N}(b^*(s^M))$$

is so to speak the *$h_G$-degree* of the map $b: B^+ \wedge S^N \to S^M$. Of course, $1_B \in h_G(B)$ denotes the unit element. A straightforward calculation shows further that *the $\pi_G$-module structure on $h_G$ is the interior multiplication by $u(-)$.*

**3.5 Corollary.** *In a multiplicative $G$-cohomology theory, the index of a $G$-fixed point situation $f$ over $B$ takes the form*

$$I_{h_G}(f) = I_G(f)\cdot 1_B = u(I_G(f))$$

*where $1_B$ denotes the unit element in $h_G(B)$. Thus, when the $\pi_G$-index $I_G(f)$ of $f$ is represented by the $G$-map $i_G(f): B^+ \wedge S^M \to S^M$, then $I_{h_G}(f)$ is its $h_G$-degree $\sigma^{M,-M}(i_G(f)^*(s^M))$.* □

**3.6** In particular, the fixed point indices in $h^0_G$ are stably spherical cohomology classes - which can be seen equally well by a geometric argument (see [Dold 2], 3.7).
Conversely, any stably spherical cohomology class of degree zero arises as a fixed point index: For, let the G-map $\psi: B \times (M, M-0) \to (M, M-0)$ be a representative for some element in $\pi^0_G(B)$. Then the vertical G-map $f: B \times M \to B \times M$, $(b,x) \mapsto (b, x - \psi(b,x))$ is compactly fixed because $\mathrm{Fix}(f) = \psi^{-1}(0)$ is contained in $B \times 0$, and the $\pi_G$-index of $f$ is represented by $\psi$ as is obvious from the very definition.

**Theorem.** *If $h_G$ is a multiplicative G-cohomology theory, then precisely the stably spherical cohomology classes in $h^0_G(B)$ occur as indices of G-fixed point situations over B.* □

**3.7** The next result emphasizes the universal character of the $\pi_G$-index.

**Proposition.** *Let $f$ be a G-fixed point situation over a base with trivial G-action. If $B$ is pathwise connected, then all fibres of $f$ have one and the same $\pi_G$-index $I_G(f_{pt}) \in \pi^0_G(\mathrm{pt})$.*
*If, furthermore, $B$ is of finite category, so for example a finite-dimensional CW-complex, then the index homomorphism of $f$ localized away from $I_G(f_{pt})$ is an isomorphism for every G-cohomology theory $h_G$. I.e.: $I^f_{h_G}$ induces an automorphism of the local module $h_G(B)\left[(I_G(f_{pt}))^{-1}\right]$ over the ring $\pi_G(\mathrm{pt})\left[(I_G(f_{pt}))^{-1}\right]$.*

PROOF. The first assertion is an immediate consequence of HTP. For the rest, we abbreviate $I_G(f_{pt})$ by $I$. Since $I^f_{h_G}$ is multiplication by the $\pi_G$-index $I^f_G$ (3.2), we only have to show that $I_G(f)$ can be inverted in the local ring $\pi_G(B)[I^{-1}]$.
By definiton, $I_G(f)$ differs from $I \cdot 1_B$ merely by some element $b$ in the kernel of $\pi_G(B) \to \pi_G(\mathrm{pt}) : I_G(f) = I \cdot 1_B + b$. And $(1_B + I^{-1} \cdot b)$ will be a unit in $\pi_G(B)\left[I^{-1}\right]$ when $b$ proves to be nilpotent since then, the series $\left(1_B - (I^{-1}b) - (I^{-1}b)^2 - \cdots\right)$ terminates eventually.
In fact, $b$ comes from $\pi_G(B, \mathrm{pt})$ via the inclusion $j: B \to (B, \mathrm{pt})$. The $n$-fold exterior power of a preimage of $b$ is an element $\bar{b}$ in the stable G-cohomotopy group of $(B, \mathrm{pt})^n = (B^n, \mathsf{V}^n B)$ where $\mathsf{V}$ denotes the fat wedge, and $j^n \circ \Delta_n: B \to B^n \to (B^n, \mathsf{V}^n B)$ maps $\bar{b}$ to $b^n$. If $B$ is of finite category, the diagonal $\Delta_n$ will factorize, up to homotopy, over the inclusion $i$: $\mathsf{V}^n B \to B^n$ as soon as $n$ becomes sufficiently large. Thus, $b^n$ vanishes for large $n$ because $(j^n\, i)^*$ is trivial. □

Corollary 3.5 thus implies

**3.8 Corollary.** *Let $f$ be as above and let $h_G$ be a multiplicative $G$-cohomology theory. The $h_G$-index of $f$ is an invertible element of the ring $h_G^0(B)$ if and only if the $h_G$-index of any fibre $f_{pt}$ of $f$ is an invertible element of the ring $h_G^0(\mathrm{pt})$. This in turn comes true if the $\pi_G$-index of $f_{pt}$ is invertible.* □

**3.9** We end with a discussion of the $\pi_G$-characteristic of a finite-sheeted $G$-covering $p$. Since the index behaves naturally under a change of base (PUB), we only need to consider for $p$ a universal $G$-covering with a finite number of sheets:
So, let $\underline{n}$ be the set $\{1, \dots, n\}$. The space of embeddings of $\underline{n}$ into a complex $G$-module $M$, denoted by $E_M(n)$, comes with a free action of the *symmetric group* $S(n)$. The corresponding orbit space $B_M(n)$ is the set of all subsets of $M$ consisting of $n$ elements. Now, the orbit projection $q_M(n)$: $E_M(n) \to B_M(n)$ is a $G$-$S(n)$-principal bundle which is locally trivial since its structure group $S(n)$ is finite (II, 5.2).
Taking the limit over complex $G$-modules $M$ and complex $G$-monomorphisms, we obtain from the direct system $\{q_M(n)\}$ a *universal $G$-$S(n)$-principal bundle* $q_G(n)$: $E_G(n) - B_G(n)$ (see [Hauschild 2]). If $B$ is a finite-dimensional G-space with finite orbit structure, then the classifying map of any numerable, $n$-sheeted $G$-covering of $B$ factorizes over $B_M(n)$ for $M$ sufficiently large.

**3.10** Let $p$: $E \to B$ be the $n$-sheeted covering $E_M(n) \times_{S(n)} \underline{n} \to B_M(n)$ associated with $q_M(n)$. According to II, 5.8 or 5.10, $p$ is a $G$-ENR$_B$, for $B$ is a metric space of finite dimension. Further, $p$ is a proper map since $p$ is a locally trivial fibre bundle with compact fibre. Therefore, the $\pi_G$-characteristic of $p$ is well-defined.
To calculate it, we realize $E$ in $B \times M$ as the vertical $G$-subspace of all pairs $(b, x)$ with $x \in b$. We may assume that $M$ is a unitary $G$-module. Then $b \mapsto \min\{ \|x - y\| : x, y \in b \text{ and } x \neq y \}/2$ is a $G$-function $\rho$: $B \to \mathbb{R}$, and the set of all points $(b, x) \in B \times M$ with $\mathrm{dist}(x, b) < \rho(b)$ forms an open $G$-neighbourhood $O \subset B \times M$ of $E$. For each $(b, x) \in O$, let $b_x$ denote the unique element in $b$ with the property $\|x - b_x\| < \rho(b)$. Then, $(b, x) \mapsto (b, b_x)$ is a vertical $G$-retraction of $O$ onto $E$.
So essentially, the $\pi_G$-characteristic of $p$: $E \to B$ is represented by the $G$-map

$$(b, x) \mapsto x - b_x : (O, O - E) \to (M, M - 0).$$

The adjoint $\omega_M$ of the corresponding $G$-map $B^+ \wedge S^M \to S^M$ maps some $b \in B$ to the pointed, non-equivariant transformation of $S^M$ which expands the open balls of radius $\rho(b)$ around the points in $b$ over the whole of $M$ via $x \mapsto (x - b_x) / (\rho(b) - \|x - b_x\|)$, pushing to infinity all the rest of $S^M$.

In other words, this $G$-map from $B = B_M(n)$ to $(S^M, S^M)^0$ assigns to a configuration $b \in M$ its *Pontryagin-Thom construction* $\mathrm{PT}(b)\colon S^M \to b^+ \wedge S^M$ followed by the projection onto $S^M$. The limit of $\{\omega_M\}$ taken over $M$, is known as the *equivariant group completion map* $\omega\colon \bigoplus_n B_G(n) \to \varinjlim_M (S^M, S^M)^0$ (see [Hauschild 2]).

**Propositon.** *Let $p$ be an $n$-sheeted $G$-covering of a paracompact, compactly generated $G$-space $B$ (see 1.1). If its classifying map $b(p)\colon B \to B_G(n)$ factorizes over $B_M(n) \to B_G(n)$, then $p$ is a $G$-$\mathrm{ENR}_B$ and $\chi_G(p) \in \pi_G^0(B)$ is represented by the composite of $b(p)$ with the equivariant group completion map $\omega\colon \bigoplus_n B_G(n) \to \varinjlim_M (S^M, S^M)^0$ or rather by the adjoint thereof.* □

**3.11** From the above discussion, we deduce easily that the stable $G$-transfer of a finite $G$-set $X$ is the Pontrjagin-Thom construction for a $G$-embedding of $X$ into some complex $G$-module. This holds for any compact smooth $G$-manifold $X$:

**Proposition.** *Let $X$ be a compact smooth $G$-manifold, and let $\mathrm{PT}\colon S^M \to X^N$ denote the Pontrjagin-Thom construction corresponding to a $G$-embedding of $X$ into a complex $G$-module $M$ with normal bundle $N \to X$.*
*Regarding $X$ as a $G$-ENR over a point, its stable $G$-transfer is represented by the composite of PT with the the projection $X^N \to X^+ \wedge S^M$ of Thom spaces induced by the inclusion $N \to X \times M$.*

PROOF. If $X$ has been embedded into a $G$-module $M$, then, on a sufficiently small closed $\varepsilon$-neighbourhood $\overline{N_\varepsilon}$ of the zero-section in the normal bundle $N \subset X \times M$, the addition $a\colon X \times M \to M$ is a $G$-homeomorphism with a $G$-tubular neighbourhood of $X$ in $M$. Since $a$ is the identity on the zero-section $X$, the projection $r\colon a(N_\varepsilon) \xrightarrow{\approx} N_\varepsilon \to X$ is a $G$-neighbourhood retraction in $M$ onto $X$.
Thus, in case $M$ is a complex $G$-module, the stable $G$-transfer $t_G^X = r^+ \circ t_G^{ir}$ (2.8) of the projection $X \to \mathrm{pt}$ is represented by the composite of the inclusions $(M, M-0) \xleftarrow{\simeq} (M, M-\text{ball}) \rightarrowtail (M, M-x) \xleftarrow{\simeq} (a(N_\varepsilon), a(N_\varepsilon)-X)$ with the $G$-map $(r, i-ir)\colon (a(N_\varepsilon), a(N_\varepsilon)-X) \to X \times (M, M-0)$ where $i$ is the inclusion $X \to a(N_\varepsilon) \to M$. Because of $(r, i-ir)(a(x,v)) = (x, v)$, the above is indeed the Pontrjagin-Thom construction $\mathrm{PT}(a)\colon S^M \to \overline{N_\varepsilon}/\dot{N_\varepsilon} \to X^N$, followed by the map $X^N \to X^+ \wedge S^M$ of Thom spaces which is induced by the inclusion of $N$ into $N \oplus T(X) = X \times M$ over $X$. □

**3.12** In the *non-equivariant case*, the universal $n$-sheeted covering $E(n) \times_{S(n)} \underline{n} \to B(n)$ has a characteristic in $\pi^0(B(n))$ which coincides with the above $\pi$-characteristic over the skeletons

$B^{\nu}(n)$ which in fact are finite CW-spaces (see [Clapp-Prieto]). $\pi^0(B(n))$ is now the inverse limit of the stable cohomotopy groups $\pi^0(B^{\nu}(n))$ because the inverse system $\pi^{-1}(B^{\nu}(n))$ satisfies the Mittag-Leffler condition: For, in all degrees other than zero, the stable cohomotopy groups of a finite CW-space are finite. This in turn follows by induction on cells as the stable $k$-stems are finite for $k \neq 0$.

**Corollary.** *Let $p$ be an $n$-sheeted covering of a* CW*-space $B$. Its Clapp-Prieto characteristic in stable cohomotopy is represented by (the adjoint of) the composite of its classifying map $b(p)\colon B \to B(n)$ with the group completion map $\omega\colon \bigoplus_n B(n) \to \Omega^\infty S^\infty$.*
*In particular, on $B(n)$, the group completion map is the $\pi$-characteristic of the universal covering $E(n) \times_{S(n)} \underline{n} \to B(n)$.* □

**3.13** Example. *Let $G$ be a finite group with classifying space $B(G)$.* For every *finite $G$-set $X$*, $p_G[X]\colon E(G) \times_G X \to B(G)$ is a finite-sheeted covering. Its classifying map $B(G) \to B(|X|)$ is induced by the action $G \to S(|X|)$ of $G$ on $X$. Composed with the group completion map, it represents the $\pi$-characteristic of $p_G[X]$, for $B(G)$ is a CW-space.
*The assignment $X \mapsto \chi_\pi(p_G[X])$ induces G. Segal's map from the Burnside ring $A(G)$ of $G$ to $\pi^0(B(G))$.* We will encounter an alternate description of this map in the next section (see 4.18).

## 4. The Equivariant Fixed Point Ring

**4.1** Having seen in 3.6 that all stable $G$-cohomotopy classes in $B$ of degree zero are indices of $G$-fixed point situations over $B$, we now want to provide a geometric criterion for two $G$-fixed point situations to have the same $\pi_G$-index.
As before, $G$ is assumed to be a *compact Lie group* and $B$ is a *paracompact* $G$-space whose topology is *compactly generated.*

**4.2 Definition and Proposition.** Two $G$-fixed point situations $f_0$ and $f_1$ over $B$ are said to be *equivalent,* in symbols $f_0 \sim f_1$, if there is a $G$-fixed point situation over $B \times [0, 1]$ whose parts over $B \times \{i\}$ coincide with $f_i$ for $i = 0, 1$. Indeed, *this is an equivalence relation.*
Let $Fix_G(B)$ denote the set of corresponding equivalence classes $[f]$. *Taking topological sums and fibrewise products of representatives, we equip $Fix_G(B)$ with the structure of a commutative ring with unit element $[\mathrm{id}_B]$. Its zero element is the empty class $[\emptyset]$.*

PROOF. The relation "$\sim$" is transitive: Let $f_{[0,1]}: V_{[0,1]} \to E_{[0,1]}$ and $f_{[2,3]}: V_{[2,3]} \to E_{[2,3]}$ be $G$-fixed point situations over $B \times [0, 1]$ and $B \times [2, 3]$ respectively such that the fibres $f_1$ and $f_2$ coincide. Then, $f_{[0,1]} \cup f_1 \times \mathrm{id}_{[1,2]} \cup f_{[2,3]}$ is a compactly fixed $G$-map over $B$ which connects $f_0$ to $f_3$ in $p: E_{[0,1]} \cup E_1 \times [1, 2] \cup E_{[2,3]} \to B \times [0, 3]$. By Proposition II, 1.11, $p$ is a vertical $G$-ENR, for the open subspaces $p^{-1}(B \times [0, 2))$ and $p^{-1}(B \times (1, 3])$ are both vertical $G$-ENRs over $B \times [0, 2)$ and $B \times (1, 3]$ respectively, and hence over $B \times [0, 3]$.
$Fix_G(B)$ is a commutative ring: Clearly, $f \oplus f'$ and $f \times_B f'$ are $G$-fixed point situations over $B$ if $f$ and $f'$ are so, and both of these operations respect the relation "$\sim$". The axioms of a semiring are satisfied and we claim that the identity on $B \times (S^1 \vee S^1)$ represents the additive inverse of the unit element $[\mathrm{id}_B]$. For this, take a copy of $S^1$, secure to it an elastic membrane, and embed concentrically a solid annulus. Denote the resulting disc by $D$, its boundary by $S$, and the annulus by $A$.
On the one hand then, the identity on $D \vee S$ is equivalent to the identity on $S$ via the mapping cone of the retraction $D \vee S \to S$, and gradually distorting the half-open cylinder $[0, 1) \times S$, we obtain a vertical fixed point situation over $[0, 1]$ which connects $\mathrm{id}_S$ to the empty map $\emptyset$.
On the other hand, $[0, 1] \times (D \vee S) - \{1\} \times A$, being an open subspace of $[0, 1] \times (D \vee S)$, is a vertical ENR over $[0, 1]$ and gradually distorting the annulus $A$, we deform the identity on $D \vee S$ to a fixed point situation $d$ in $(D - A) \vee S$ which keeps fixed $S \vee S$ and the centre $P$ of $D$ and nothing else. $d$ is equivalent to the identity on $P \oplus (S \vee S)$: To realize this, we erect over $(D - \mathring{A}) \vee S$ the mapping cylinder of the retraction onto $P \oplus (S \vee S)$ and remove from its bottom the boundary of $A$. The result is a vertical ENR $p$ over $[0, 1]$ and we merely let propagate the distortion $d$ from its bottom $p_0$ to each of its layers $p_t$.
To summarize, the identity on $P \oplus (S \vee S)$ is equivalent to the empty map. Taking the product with $B$, we have that the identity on $B \oplus B \times (S^1 \vee S^1)$ represents the zero element $[\emptyset]$ in $Fix_G(B)$. □

**4.3** Due to its homotopy invariance, the index defines a map $I_{h_G}: Fix_G(B) \to h_G^0(B)$ for any multiplicative $G$-cohomology theory $h_G$. And the properties ADD, Mul, and IDT just say that $I_{h_G}$ is a homomorphism of unitary rings.
According to Theorem 3.6, its image consists of the stably spherical cohomology classes in $B$. In particular, $I_G = I_{\pi_G}$ is an epimorphism and $I_{h_G}$ decomposes into $u \circ I_G$ (see 3.4).

**Theorem.** *Let $h_G$ be a multiplicative $G$-cohomology theory, and let $u$ denote the corresponding degree map (3.4). Then*

$$\begin{array}{ccc} & Fix_G(B) & \\ {}^{I_G}\swarrow\,\cong & & \searrow^{I_{h_G}} \\ \pi^0_G(B) & \xrightarrow{\quad u \quad} & h^0_G(B) \end{array}$$

*is a communicative diagram of unitary ring homomorphisms wherein $I_G$ is an isomorphism.*

PROOF. It remains to show that $I_G$ is injective. For this, we follow [Dold 2], Section 4, to transform any vertical $G$-fixed point situation into a representative of simple shape.

**4.4 Lemma** (LOC). *Let $f_1$: $V_1 \to E_1$ be a G-fixed point situation over B which maps an open G-neighbourhood $V_0$ of its fixed point set to a G-numerically open subspace $E_0 \supset V_0$ of $E_1$. Then $f_1$ is equivalent to the induced G-fixed point situation $f_0$: $V_0 \to E_0$.*

PROOF. Let $f$: $V \to E$ denote the union

$$f_0 \times \mathrm{id} \cup f_1 \times \mathrm{id}\colon\ V_0 \times [0, 1] \cup V_1 \times (0, 1] \longrightarrow E_0 \times [0, 1] \cup E_1 \times (0, 1].$$

As a $G$-numerically open subspace of $E_1 \times [0, 1]$, $E$ is a $G$-ENR over $B \times [0, 1]$ (II, 1.8). $V$ is open in $E$ and $f$ is a compactly fixed $G$-map over $B \times [0, 1]$ - its fixed point set is $\mathrm{Fix}(f_1) \times [0, 1]$ - which connects $f_0$ to $f_1$. □

**4.5** A closed inclusion $i$: $D \to E$ of $G$-ENR$_B$s is a $G$-cofibration over $B$ (II, 1.9). Therefore, by a theorem of Strøm (see [tom Dieck-Kamps-Puppe], 1.22), $Z := E \times \{0\} \cup D \times [0, 1]$ is a vertical $G$-retract of $E \times [0, 1]$ over $B$ and hence a $G$-ENR$_B$ (II, 1.8). By the way, $Z$ is the mapping cylinder of $i$ because $i$ is a closed inclusion.

**Lemma** (SHR). *With the above notation, let $f$: $U \to Z$ be a G-fixed point situation over B ranging entirely either in $E = E \times \{0\} \subset Z$ or in $D = D \times \{1\} \subset Z$. Then f is equivalent to the induced G-fixed point situation $f_0$: $U \cap E \to E$ or $f_1$: $U \cap D \to D$ respectively.*

PROOF. The wedge $W := \{(x, t, s) \in E \times [0, 1] \times [0, 1]\colon (x, t) \in Z \text{ and } s \geq t\}$, being a vertical $G$-retract of $Z \times [0, 1]$,, is a $G$-ENR over $B \times [0, 1]$, which connects $f$ to $f_0$. For $f_1$, take $s \leq t$ in $W$. □

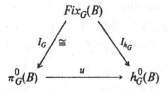

**4.6 Lemma** (COM). *Let $r: E \to D$ be a vertical G-retraction of a G-ENR$_B$ $p: E \to B$ onto a G-subspace $i: D \to E$, i.e. $ri = \mathrm{id}_D$. Then every G-fixed point situation $f: U \to D$ in $p|D$ is equivalent to the G-fixed point situation $ifr: r^{-1}(U) \to E$ in p.*

PROOF. With $Z$ as above, the map $\left(r^{-1}(U) \times \{0\} \cup U \times [0,1]_t\right) \times [0,1]_s \to Z \times [0,1]_s$, $(x,t,s) \mapsto (ifr(x), s, s)$ is a $G$-fixed point situation over $B \times [0,1]_s$. According to the SHR-lemma, its parts over $B \times \{0\}$ and $B \times \{1\}$ are equivalent to $ifr$ and $f$ respectively. □

**4.7** Like in [ Dold 2 ], 3.6, we now get the following intermediate result:

**Lemma.** *Let $f: V \to E$ be any G-fixed point situation over B. There exists a G-fixed point situation $f': E \times \mathbb{R} \to E \times \mathbb{R}$ over B with fixed point set* $\mathrm{Fix}(f) \times \{0\}$ *such that, in some G-neighbourhood $U \subset V$ of* $\mathrm{Fix}(f)$, *$f'(u,t)$ equals $(f(u), 0)$ for all $t \in \mathbb{R}$.*
According to the COM-lemma, $f'|(U \times \mathbb{R})$ is equivalent to $f|U$. Thus, by excision (4.4), *we have $[f'] = [f]$ in $Fix_G(B)$.* □

**4.8 Proposition.** *In every class of $Fix_G(B)$, there exists a globally defined G-fixed point situation of the form $f: B \times M \to B \times M$, $(b,x) \mapsto (b, \varphi(b,x))$ with M a complex G-module which has the properties*

$$\|x - \varphi(b,x)\| = \|x\| \quad \textit{and} \quad \varphi(b, tx) = t\varphi(b,x).$$

*for $(b,x) \in B \times M$ and $t \in [0,\infty)$.*

PROOF. With the lemmata provided, the proof proceeds as in [Dold 2], 4.8. If $M$ fails to be a complex $G$-module, yet, we replace $f$ by its product with the constant homomorphism on $M$: According to the COM-lemma, $f \times 0$ is a $G$-fixed point situation in $B \times M \times M$ over $B$ equivalent to $f$. It still enjoys both of the properties stated above, and now $M \times M$ is a complex $G$-module. □

**4.9** Since for a $G$-fixed point situation $f$ as in 4.8, the difference $i - f: B \times M \to B \times M$, $(b,x) \mapsto (b, x - \varphi(b,x))$ is an isometric map, Definition 1.4 implies immediately:

**Corollary.** *Let $f: B \times M \to B \times M$ be a G-fixed point situation as in 4.8. Then its stable G-index morphism $i_G^f: B^+ \to B^+$ is represented by the G-map $i - f$: $B \times (M, M - 0) \to B \times (M, M - 0)$, or by the corresponding G-transformation on the Thom space $B^+ \wedge S^M$.*

*Hence, $I^f_{h_G}$ is the M-fold desuspension of $h_G(i-f)$. And the $\pi_G$-index of $f$ has as representative the G-map* $\mathrm{proj}_M \circ (i-f)\colon B \times (M, M-0) \to B \times (M, M-0)$, $(b, x) \mapsto x - \varphi(b, x)$, *or the induced G-map* $i_G(f)\colon B^+ \wedge S^M \to S^M$ *of Thom spaces.* □

4.10 PROOF OF THEOREM 4.3. It remains to show that $I_G\colon Fix_G(B) \to \pi^0_G(B)$ is injective.
Let $f_i = (\mathrm{proj}_B, \varphi_i)\colon B \times M_i \to B \times M_i$, $i = 0, 1$, be vertical $G$-fixed point situations as in Proposition 4.8 having the same $\pi_G$-index. That is, the corresponding maps $i_G(f_i)\colon B^+ \wedge S^{M_i} \to S^{M_i}$ from 4.9 are to be homotopic as pointed stable $G$-maps. Then there exist (complex) $G$-modules $N_0$ and $N_1$ such that, up to a $G$-isomorphism $M_0 \times N_0 \cong M_1 \times N_1$, the $N_0$-fold suspension of $i_G(f_0)$ is connected to the $N_1$-fold suspension of $i_G(f_1)$ through a pointed $G$-homotopy. We clearly may assume $M_0 \times N_0 = M_1 \times N_1 = L$. So, let $\phi_t\colon B^+ \wedge S^L \to S^L$ be a pointed $G$-homotopy leading from $(i_G(f_0) \wedge \mathrm{id}_{S^{N_0}})$ to $(i_G(f_1) \wedge \mathrm{id}_{S^{N_1}})$.
When we think of $S^L$ as $L \cup \{\infty\}$, then $E := [0, 1] \times B \times L$ is a $G$-subspace of $[0, 1] \times (B^+ \wedge S^L)$. $V := \phi^{-1}(L)$ is an open $G$-subspace of $E$ because $\phi$ is a pointed $G$-homotopy, and $F\colon V \to E$, $(t, b, x) \mapsto (t, b, x - \phi_t(b, x))$ is a $G$-fixed point situation over $[0, 1] \times B$ because $\mathrm{Fix}(F) = \phi^{-1}(0)$ lies properly over $[0, 1] \times B$: For, if $A \subset B$ is compact, then $\phi^{-1}(0) \cap ([0, 1] \times A \times L) = \phi^{-1}(0) \cap ([0, 1] \times (A^+ \wedge S^L))$ is compact as well.
The parts of $F$ over $\{0\} \times B$ and $\{1\} \times B$ are the vertical $G$-fixed situations $f_0 \times 0$ and $f_1 \times 1$ in $(B \times M_0) \times N_0 = (B \times M_1) \times N_1$ which are equivalent to $f_0$ and $f_1$ respectively because of the COM-lemma. □

4.11 In order to realize in $B$ stable $G$-cohomotopy classes of non-trivial degree $\alpha$, we consider $G$-fixed point situations of degree $\alpha$ over $B$. Remember that $\alpha$ is a virtual representation of $G$ over $\mathbb{R}$. In a unique way, $\alpha$ writes as the difference of real $G$-modules $L$ and $N$ which have no non-trivial $G$-submodules in common.

**Definition and Proposition.** Let $\alpha = N - L$ be a virtual representation of $G$ over $\mathbb{R}$ with disjoint $G$-modules $L$ and $N$ and let $p\colon E \to B$ be a $G$-$\mathrm{ENR}_B$.
By a *$G$-fixed point situation of degree $\alpha$ in $p$*, we understand a vertical $G$-map $f\colon V \times L \to E \times N$, with $V \subset E$ an open $G$-subspace, whose coincidence set with the map $\imath\colon V \times L \to E \times N$, $(v, x) \mapsto (v, 0)$ lies properly over $B$. By abuse, we write $\mathrm{Fix}(f)$ for the coincidence set $\{(v, x) \in V \times L : f(v, x) = (v, 0)\}$.

*Let $r$ and $i$ be the data of a vertical $G$-neighbourhood retraction onto $p$ in a vertically trivial $G$-$\mathrm{ENR}_B$ $B \times M \to B$, i.e. $ri = \mathrm{id}_E$. Then, $f' := (i \times \mathrm{id}_N) f (r \times \mathrm{id}_L)\colon V' \times L \to (B \times M) \times N$*

*with* $V' := r^{-1}(V) \subset B \times M$ *is a G-fixed point situation over B and like in 1.4, it yields a stable G-morphism*

$$B \times (M \times L, M \times L - 0) \longrightarrow (V' \times L, V' \times L - \mathrm{Fix}(f')) \xrightarrow{\iota' - f'} B \times (M \times N, M \times N - 0).$$

*Its* $(M \oplus L)$*-fold suspension, a morphism in* $\underline{\mathrm{Stab}}_G$ *from* $B^+$ *to* $(B^+, \alpha)$*, depends only on f.* As before, we call it the *stable G-index morphism of f,* denoted by $i_G^f$. □

**4.12** If $f$ is a $G$-fixed point of degree $\alpha$ over $B$, then, in any $G$-cohomology theory $h_G$, the stable $G$-index morphism of $f$ induces an endomorphism $I_{h_G}^f$ of the graded group $h_G^\beta(B)$ *of degree* $\alpha$. Again, $I_{h_G}^f$ is a homomorphism of $h_G(B)$-modules when $h_G$ is a multiplicative theory.

**Proposition and Definition.** *Let* $h_G$ *be a multiplicative G-cohomology theory. The index homomorphism of a G-fixed point situation f of degree* $\alpha$ *over B is the multiplication by the image of the unit element* $1_B \in h_G^0(B)$.
As before, we call it the $h_G$*-index of f,* denoted by $I_{h_G}(f) \in h_G^\alpha(B)$. *Its properties are completely analogous to those of the fixed point index in degree* 0. □

**4.13** It is obvious how we have to declare the *graded ring* $Fix_G^*$ (cf 4.2): It consists of equivalence classes of $G$-fixed point situations of degree $\alpha$ over $B$. Thus, $Fix_G^0(B)$ agrees with $Fix_G(B)$. As one might expect, elements of degree $\alpha = N - L$ in $Fix_G^*(B)$ are represented by globally defined $G$-fixed point situations of the form $f: (B \times M) \times L \to (B \times M) \times N$ over $B$ for which $\iota - f$ is an isometric map, thereby inducing a $G$-map $B^+ \wedge S^{M \oplus L} \to B^+ \wedge S^{M \oplus L}$. As before, we get:

**Theorem.** *The index* $I_G$: $Fix_G^*(B) \to \pi_G^*(B)$ *is an isomorphism of graded rings. When* $h_G$ *is a multiplicative G-cohomology theory, precisely the stably spherical cohomology classes in* $h_G^*(B)$ *occur as indices of G-fixed point situations of degree* $\alpha$ *over B.* □

**4.14** COMMENT. In fact, *the contravariant functors* $Fix_G^*$ *constitute an equivariant cohomology theory naturally isomorphic to equivariant stable cohomotopy theory by virtue of the* $\pi_G$*-index* $I_G$. For the non-equivariant case, this has been detailed in [Prieto].

**4.15** We end by interpreting various maps between equivariant stable cohomotopy groups in terms of equivariant fixed point theory.

We already know that an induced map $\pi_G(b\colon B' \to B)$ is reflected by taking the pull-back $b^*(-)$ (PUB), and that the $\pi_G$-transfer of a proper $G$-ENR$_B$ $q\colon C \to B$ turns a $G$-fixed point situation over $C$ into a $G$-fixed point situation $q_*(-)$ over $B$ (2.11). Further, the structure map $\theta^\lambda\colon \pi_G \to \pi_H^{\lambda^*(\alpha)}$, with $\lambda\colon H \to G$ a Lie homomorphism, makes us look at a $G$- as an $H$-fixed point situation (NAT).

For an $H$-space $B$ whereon the kernel of $\lambda$ acts freely, the composite

$$\pi_G^\alpha(G \times_H B) \xrightarrow{\theta^\lambda} \pi_H^{\lambda^*(\alpha)}(G \times_H B) \xrightarrow{(\lambda \times_H \mathrm{id}_B)^*} \pi_H^{\lambda^*(\alpha)}(B)$$

is an isomorphism (see 2.2). PUB and NAT therefore imply:

**Proposition.** *Let $\lambda\colon H \to G$ be a Lie homomorphism and suppose $B$ is an $H$-space on which the kernel of $\lambda$ acts freely. Pulling back to $B$ any $G$-fixed point situation over $G \times_H B$ is an isomorphism*

$$Fix_G^\alpha(G \times_H B) \xrightarrow{\cong} Fix_H^{\lambda^*(\alpha)}(B)$$

*in every degree $\alpha$.*

*In particular, $Fix_G^*(G/H)$ is isomorphic to $Fix_H^{\lambda^*(\alpha)}(\mathrm{pt})$ for any subgroup $H \leq G$. In fact, a $G$-fixed point situation over $G/H$ is uniquely determined its fibre over the coset $H$.* □

**4.16 Corollary.** *With the same assumptions as before, let $p$ be an $H$-ENR$_B$ for which $\mathrm{id}_G \times_H p$ is a $G$-ENR over $G \times_H B$. For example, $\lambda\colon H \to G$ might be the inclusion of a subgroup (II, 6.2).*

*Then, for any $H$-fixed point situation $f$ in $p$ of degree $\lambda^*(\alpha)$, $\mathrm{id}_G \times_H f$ is a $G$-fixed point situation in $\mathrm{id}_G \times_H p$ of degree $\alpha$. In an equivariant cohomology theory defined on $\lambda$, their fixed point indices correspond under the isomorphism $(\lambda \times_H \mathrm{id}_B)^* \circ \theta^\lambda\colon h_G^*(G \times_H B) \xrightarrow{\cong} h_H^{\lambda^*(\alpha)}(B)$ (see 2.2).*

*For $K \leq H \leq G$, in particular, the $h_G$-characteristic of the bundle $G/K \to G/H$ corresponds to the $h_H$-characteristic of its fibre $H/K$.*

PROOF. Let $f = (\varphi, \psi)\colon V \times \lambda^*(L) \to E \times \lambda^*(N)$ be an $H$-fixed point situation in $p$ of degree $\lambda^*(\alpha)$, $\alpha = N - L$. $(g, v, x) \mapsto (g, v, gx)$ induces a vertical $G$-homeomorphism of $\mathrm{id}_G \times_H f$ with the map

$$([g, v], x) \mapsto ([g, \varphi(v, g^{-1}x)], g\psi(v, g^{-1}x))\colon (G \times_H V) \times L \to (G \times_H E) \times N.$$

And this is a $G$-fixed point situation in $\mathrm{id}_G \times_H B$ of degree $\alpha$ since the coincidence set $G \times_H \mathrm{Fix}(f)$ of $\mathrm{id}_G \times_H f$ with $\mathrm{id}_G \times_H \imath$ lies properly over $G \times_H B$. Pulling it back to $B$, we get back $f$. □

**4.17** Interchange $H$ and $G$ and consider the constant homomorphism $\lambda = e : G \to H$.

**Corollary.** *Let $p$ be a $G$-$\mathrm{ENR}_B$ over a free $G$-space $B$ such that $p/G$ is an* ENR *over $B/G$. If $f$ is a $G$-fixed point situation in $p$ of degree $n \in \mathbb{Z}$, then the $h_G$-index of $f$ and the $h$-index of $f/G$ correspond in any multiplicative equivariant cohomology theory defined on $e: G \to \{e\}$.*
□

**4.18** Example. Let $G$ be a *finite group* and let $E(G) \to B(G)$ denote its classifying bundle. For every finite $G$-set $X$, $p[X]: E(G) \times X \to B(G)$ is then a vertical $G$-ENR over a free $G$-space, and $p[X]/G$ is the finite-sheeted covering $p_G[X]: E(G) \times_G X \to B(G)$ from 3.11. Therefore, *the Segal map $[X] \mapsto \chi_\pi(p_G[X])$ from the Burnside ring $A(G)$ of $G$ to $\pi_G^0(B(G))$ is equivalent to the map $[X] \mapsto \chi_G(p[X]) : A(G) \to \pi_G^0(E(G))$.*

**4.19** Remark. In equivariant stable cohomotopy theory and for $\lambda = \imath : H \leq G$, the *inverse $\eta_\imath$ of the natural isomorphism $(\imath \times_H \mathrm{id}_B) \circ \theta^\imath : \pi_G^0(G \times_H B) \to \pi_H^0(B)$ has the following geometric interpretation:*
Since any $H$-module is a direct summand in some $G$-module, every element of $\pi_H^0(B)$ can be represented by a pointed $H$-map $\psi: B^+ \wedge S^M \to S^M$ with $M$ a (complex) $G$-module. In the pointed category, $H$-maps from an $H$-space $Z$ to some $G$-space $Y$ correspond one-to-one to $G$-maps $G^+ \wedge_H Z \to Y$, and for $X$ a $G$-space, $G^+ \wedge_H (B^+ \wedge X)$ is $G$-homeomorphic to $(G^+ \wedge_H B^+) \wedge X$ by virtue of $[g,(b,x)] \mapsto ([g,b], gx)$.
So, to our $H$-map $\psi$, there corresponds the $G$-map

$$\varphi : (G \times_H B)^+ \wedge S^M \to S^M, \ \ ([g,b],x) \mapsto g\psi(b, g^{-1}x).$$

Indeed, when we regard $\varphi$ as an $H$-map and restrict it to $B^+ \wedge S^M$, we get back $\psi$.

This shows further that *for a $G$-space $B$, the composite*

$$\pi_G^0(B) \xrightarrow{\theta^\imath} \pi_H^0(B) \xrightarrow{\eta_\imath} \pi_G^0(G \times_H B) \cong \pi_G^0(G/H \times B)$$

*is induced by the projection $G/H \times B \to B$.* □

**4.20** REMARK. In *equivariant K-theory*, there is a homomorphism

$$\eta_\lambda : K_H(B) \to K_G(G \times_H B), \ [p] \mapsto [\mathrm{id}_G \times_H p]$$

for any $H$-space $B$ and any $\lambda: H \to G$. In fact, $\eta_\lambda$ *is the inverse of the structure isomorphism* $(\lambda \times_H \mathrm{id}_B)^* \circ \theta_\lambda$ in case the kernel of $\lambda$ acts freely on $B$ (see 2.2).
For the constant homomorphismus $e$, in particular,

$$\eta_e : K_H(B) \to K(B/H) , \ [p] \mapsto [p/H]$$

*is an isomorphism if $B$ is a free $H$-space, and for $i: H \leq G$ and any $G$-space $B$, the composite*

$$K_G(B) \xrightarrow{\theta^i} K_H(B) \xrightarrow{\eta_i} K_G(G \times_H B) \cong K_G(G/H \times B)$$

*is induced by the projection $G/H \times B \to B$.* □

**4.21** REMARK. *Suppose $B$ is an $H$-space for some $H \leq G$ such that the orbit projection $p: G \times_H B \to B/H$ is a vertical $G$-ENR.*
Locally, there is then only one orbit type on $B$ and conversely, this condition will make $p$ a vertical $G$-ENR when $B$ is in addition (see 4.1) perfectly normal and finite-dimensional, or just compact (II, 5.11).
Then, given any equivariant cohomology theory $h_{\underline{G}}$, the composite of the transfer of $p$ with the natural isomorphism $\eta_i$ is what we call *induction (with respect to $p$):*

$$\mathrm{ind} = \mathrm{ind}^p_{h_{\underline{G}}} : h^0_H(B) \xrightarrow{\eta_i} h^0_G(G \times_H B) \xrightarrow{T^p_{h_G}} h^0_G(B/H) .$$

When $B$ is a trivial $H$-space, then $p$ is the projection $G/H \times B \to B$ and the induction becomes a homomorphism $h^0_H(B) \to h^0_G(B)$. In equivariant $K$-theory, in particular, the induction for $B$ a point is a homomorphism

$$\mathrm{ind}^G_H \colon R(H) \to R(G)$$

of complex representation rings. In fact, we will see in the last section that $\mathrm{ind}^G_H$ *assigns to an $H$-module the induced representation of $G$.*

The remarks in 4.19 and 4.20 show finally that *for B a G-space, the composite*

$$h_G^0(B/H) \longrightarrow h_G^0(B) \xrightarrow{\theta^i} h_H^0(B) \xrightarrow{\mathrm{ind}_{hG}^p} h_H^0(B/H)$$

*is the index homomorphism of* $p\colon G\times_H B \to B/H$ *when* $h_{\underline{G}}$ *is equivariant stable cohomotopy theory or equivariant K-theory.*

**4.22 Corollary.** *Let* $p\colon G\times_H B\to B/H$ *be as above (4.21). Interpreted in equivariant fixed point theory, the induction*

$$\mathrm{ind}^p : \pi_H^0(B) \to \pi_G^0(B/H)$$

*assigns to some H-fixed point situation f over B the map* $\mathrm{id}_G \times_H f$ *regarded as a G-fixed point situation over* $B/H$ *by means of p.*
*Hence, the induction in a multiplicative equivariant cohomology theory* $h_{\underline{G}}$ *maps the* $h_H$*-index of f (over B) to the* $h_G$*-index of* $\mathrm{id}_G\times_H f$ *(over* $B/H$*). In particular,* $\mathrm{ind}^p(1_B)$ *is the* $h_G$*-characteristic of p.* □

## 5. A Sum Formula

**5.1** We want to extend the central theorem 5.3.13 in [tom Dieck 2] to equivariant fixed point indices.
Throughout, $G$ is assumed to be a *compact Lie group* and $h_G$ denotes a *multiplicative* $G$-cohomology theory.

**5.2 Lemma.** *Let f be a G-fixed point situation in a* $G\text{-ENR}_B$ $p\colon E\to B$ *with trivial G-action on its base. Suppose* $\emptyset \neq D \subset E$ *is a closed* $G\text{-ENR}_B$ *invariant under f.*
*Then the partial G-transformation* $f^D$ *induced by f on D is a G-fixed point situation over B and* $I_{h_G}(f) - I_{h_G}(f^D)$ *is the index of the following vertical G-fixed point situation* $f_D$ *in* $E - D$*: On a suitable G-neighbourhood* $V\subset E$ *of* $\mathrm{Fix}(f)$*, there exists a G-deformation* $d_t$ *which, at the time* $t=1$*, retracts a G-neighbourhood onto* $V\cap D$*, and which is stationary on* $V\cap D$ *and outside some G-neighbourhood of* $V\cap D$ *which lies properly over B. The restriction of* $f d_1$ *to*

*$V-D$ is then a vertical G-fixed point situation in p with $h_G$-index $I_{h_G}(f)-I_{h_G}(f^D)$. Denoting by $f_D$ its shrinking to $E-D$, we get*

$$I_{h_G}(f)-I_{h_G}(f^D) = I_{h_G}(f_D).$$

PROOF. By localizing, we restrict the domain of $f$ to a $G$-numerically open subspace $V\subset E$ whose closure lies properly over $B$. Then $V$ and $V\cap D$ are $G$-ENR$_B$s (II, 1.8) and Proposition II, 1.9 provides à vertical $G$-deformation $d_t$ of $V$ relative to $V\cap D$ such that $d_1^{-1}(V\cap D)$ is a $G$-neighbourhood of $V\cap D$.
In order to make $d_t$ stationary outside some $G$-neighbourhood of $V\cap D$, we choose a $G$-neighbourhood $U$ of $V\cap D$ in $d_1^{-1}(V\cap D)$ with closure still in $V$, and an open $G$-neighbourhood $W$ of $\overline{U}$ in $V$. Then we take a $G$-function $\tau$ on $V$ separating $\overline{U}$ and $V-W$, say $\tau(\overline{U})=1$ and $\tau(V-W)=0$, and modulate $d_t$ by means of $\tau$ setting $(v,t)\mapsto d(v,\tau(v)t)$. $\overline{W}\subset\overline{V}$ lies properly over $B$ and we may assume $\overline{W}\subset V$.
$fd_t\colon V\to E$ is a vertical $G$-homotopy of $f$ relative to $(V\cap D)\cup(V-W)$. Therefore, its fixed point set is a closed subspace of $(\overline{W}\times[0,1])\cup(\mathrm{Fix}(f)\times[0,1])$ in $V\times[0,1]$ and since $\overline{W}$ lies properly over $B$, the homotopy is compactly fixed over $B\times[0,1]$. Thus, HTP yields $I_{h_G}(f) = I_{h_G}(fd_1)$. And by ADD, we have $I_{h_G}(fd_1) = I_{h_G}(fd_1\,|\,U)+I_{h_G}(fd_1\,|\,V-D)$ because the fixed points of $fd_1$ either lie inside $V\cap D$ or outside $U$. Finally, the restriction of $fd_1$ to $U$, ranging in the $G$-ENR$_B$ $D$, has the same index as $f^D\colon V\cap D\to D$ because the latter is just its shrinking $U\cap D\xrightarrow{d_1} V\cap D\xrightarrow{f} D$ to $D$ (SHR). □

**5.3 Corollary.** *Let $f$ and $f'$ be G-fixed point situations in G-ENRs $E$ and $E'$ respectively. Suppose $D\subset E$ is a closed G-ENR invariant under $f$ and $\psi\colon D\to E'$ is a proper G-map such that $\psi f^D=f'\psi$. Then, $f\cup_\psi f'$ is a G-fixed point situation in $E\cup_\psi E'$ which has the $h_G$-index*

$$I_{h_G}(f\cup_\psi f')-I_{h_G}(f') = I_{h_G}(f)-I_{h_G}(f^D).$$

*If $D$ is a compact G-ENR, we get in particular*

$$I_{h_G}(f/D)-1_{h_G} = I_{h_G}(f)-I_{h_G}(f^D).$$

PROOF. By Proposition II, 6.5, $E\cup_\psi E'$ is a $G$-ENR and because of $\psi\circ f^D=f'\circ\psi$, $f\cup_\psi f'$ is a well-defined $G$-map whose fixed point set $\mathrm{Fix}(f)\cup_\psi\mathrm{Fix}(f')$ is compact.
Consider the $G$-deformation $d_t$ of $V$ from lemma 5.2 used to calculate $I_{h_G}(f)-I_{h_G}(f^D)$: On $V\cup_\psi V'$, it induces the $G$-deformation $d_t\cup_\psi \mathrm{id}_{V'}$ which is stationary on $V'$, and according to

Lemma 5.2, $I_{h_G}(f \cup_\psi f') - I_{h_G}(f')$ is the index of the restriction $F:(V \cup_\psi V') - V' \to E \cup_\psi E'$ of $(f \cup_\psi f') \circ (d_1 \cup_\psi \mathrm{id}_{V'}) = f d_1 \cup_\psi f'$, or of the shrinking of $F$ to $(E \cup_\psi E') - E'$. But since $F$ agrees with $f d_1 | (V - D)$ followed by the projection $E \to E \oplus E' \to E \cup_\psi E'$, its shrinking to $(E \cup_\psi E') - E'$ is just the shrinking of $f d_1$ to $E - D$. Hence, the $h_G$-index of $F$ is $I_{h_G}(f) - I_{h_G}(f^D)$. □

**5.4** As above, let $p: E \to B$ be a $G$-ENR$_B$ over a $G$-trivial base. We consider the filtration II, 6.8 of $p$ by closed $G$-ENR$_B$s $\emptyset = p_0 \subset p_1 \subset \ldots \subset p_r = p$ such that $E_i - E_{i-1} = E_{(H_i)}$, where $(H_1) \succ (H_2) \succ \ldots \succ (H_r)$ runs through the set of orbit types on $E$. Since $p^{(H_i)}$ and $p^{\underline{(H_i)}}$ are closed $G$-ENR$_B$s in $p$ (II, 6.7), a $G$-fixed point situation $f$ in $p$ induces the four vertical $G$-fixed point situations

$$\begin{array}{ccc} (V_i, V_{i-1}) & \xrightarrow{(f_i,\, f_{i-1})} & (E_i, E_{i-1}) \\ \cup & & \cup \\ \left(V^{(H_i)}, V^{\underline{(H_i)}}\right) & \xrightarrow{\left(f^{(H_i)},\, f^{\underline{(H_i)}}\right)} & \left(E^{(H_i)}, E^{\underline{(H_i)}}\right). \end{array}$$

According to Lemma 5.2, $I_{h_G}(f_i) - I_{h_G}(f_{i-1})$ is the $h_G$-index of a vertical $G$-fixed point situation, $f_{(H_i)}$ say, in $E_{(H_i)} = E_i - E_{i-1} = E^{(H_i)} - E^{\underline{(H_i)}}$. By restriction, the corresponding vertical $G$-deformation $d_t$ in $V_i$ relative to $V_{i-1}$ provides a vertical $G$-deformation of $V^{(H_i)}$ relative to $V^{\underline{(H_i)}}$.

Therefore, $I_{h_G}(f^{(H_i)}) - I_{h_G}(f^{\underline{(H_i)}})$ is the $h_G$-index of $(f d_1)^{(H_i)}$ restricted to $V_{(H_i)}$, and the shrinking of this map to $E_{(H_i)}$ is what we have called $f_{(H_i)}$. Thus, $I_{h_G}(f^{(H_i)}) - I_{h_G}(f^{\underline{(H_i)}})$ equals $I_{h_G}(f_i) - I_{h_G}(f_{i-1})$, and $I_{h_G}(f) = \sum_{i=1}^{r} \left(I_{h_G}(f_i) - I_{h_G}(f_{i-1})\right)$ shows:

**Theorem.** *The $h_G$-index of a $G$-fixed point situation $f$ over a base with trivial $G$-action decomposes into the finite sum*

$$I_{h_G}(f) = \sum_{(H)} I_{h_G}(f^{(H)}) - I_{h_G}(f^{\underline{(H)}})$$

*taken over the set of orbit types around* Fix$(f)$. □

**5.5** Remark. *The index of a $G$-fixed point situation of non trivial degree $\alpha$ decomposes the same way.*

**5.6** The above sum formula is particularly convincing if we think of the $(H)^{\text{th}}$ summand as counting the fixed points of $f$ on orbits of type $(G/H)$. However, this is not the right way: For, we have interpreted $I_{h_G}(f^{(H)}) - I_{h_G}(f^{\underline{(H)}})$ as the index of a $G$-fixed point situation $f_{(H)}$ in $E_{(H)}$ homotopic to the restriction of $f$ to $E_{(H)}$ in the ordinary sense. And this homotopy may produce additional fixed points as illustrated by the following

EXAMPLE. Let $E$ be the euclidean space $\mathbb{R}^n$ with the antipodal action of the group $G = \mathbb{Z}/2$ and let $f: E \to E$ be the stretching $x \mapsto 2x$. Then $f$ has no fixed points of type $(G/H) = G$. However, if we deform $E$ using the pointed $G$-map $d_1(x) := 2x(\|x\| - 1)/\|x\|$ for $1 \le \|x\| \le 2$, which contracts the unit ball to the origin and keeps fixed all points of norm $\|x\| \ge 2$, then, for $f_{(H)}$ with $(H) = \{e\}$, we get a $G$-fixed point situation in $E - \{0\}$ whose fixed point set is the sphere of radius 4/3.
In stable $G$-cohomotopy, we can calculate the index of $f$ by means of the monomorphism $[\varphi] \mapsto (\deg(\varphi^H)): \pi^0_G(\text{pt}) \to \prod_{H \le G} \mathbb{Z}$ (see [tom Dieck 2], 8.4.1), for by NAT, the $\pi_G$-index of $f$ goes to the family of Hopf indices $I(f^H)$ (cf 8.7).
In our example, $I_G(f)$ has the components $I(f) = (-1)^n$ and $I(f^G) = 1$, and $I_G(f^G)$ is (1, 1). For $(H) = \{e\}$ hence, $I_G(f_{(H)}) = I_G(f) - I_G(f^G)$ becomes $((-1)^n - 1, 0)$ despite the fact that $f$ has no fixed points in $E_{(H)} = E - \{0\}$.

**5.7** COMMENT. The example demonstrates that *a fixed point of $f$ in $E^K$ with $K > H$ may be counted by $f^K$ with a multiplicity different from that counted by $f^H$*. We will resume this discussion in Section 7.9.

**5.8** Over a point, we can calculate the terms in the sum formula since each of them is the index of a $G$-fixed point situation in a $G$-ENR with a single orbit type. We rely on the following

**Lemma.** *Let $f: V \to E$ be a $G$-fixed point situation in a $G$-ENR with free $G$-action. Then the $h_G$-index of $f$ is an integral multiple of the $h_G$-characteristic of $G$:*

$$I_{h_G}(f) = n(f) \cdot \chi_{h_G}(G).$$

*The weight $n(f) \in \mathbb{Z}$ does not depend on the theory $h_G$. In particular, we have*

$$\chi(G) \cdot I_{h_G}(f) = I(f) \cdot \chi_{h_G}(G)$$

*where $I$ and $\chi$ denote the Hopf index and the Euler-Poincaré characteristic calculated in ordinary singular cohomology. Further, $I_{h_G}(f)$ vanishes unless $G$ is finite.*

PROOF. Because of the universality of the $\pi_G$-index, we only have to show that $I_G(f)$ is an integral multiple of $\chi_G(G)$.

First, $\chi_G(G)$ vanishes if $G$ is not finite: For, if $g \in G$ is a non-trivial element in the component of the unit element $e$, then right multiplication by $g$ is a fixed point free $G$-transformation of $G$ which is $G$-homotopic to the identity. We will see in a moment why this implies $I_G(f) = 0$ as well.

Let $f^*: V^* \to E^*$ denote the map of orbit spaces induced by $f$. To calculate $I_G(f)$, we exploit that $p: E \to E^*$ is a locally trivial $G$-principal bundle (I, 2.11). Even more, the commutativity of the index allows us to assume that $p$ is a smooth bundle: For, $E$ is a $G$-retract of an open $G$-neighbourhood $O$ in some $G$-module and $O$ is forced to be a free $G$-space when $G$ acts freely on $E$.

Since $\mathrm{Fix}(f) \subset V$ is compact, we can arrange by excision that $\overline{V}$ is compact and that $f$ is defined on the boundary $\dot{V}$ of $V$ - without fixed points, of course. The orbit space $V^*$ is open in $E^*$ and its closure is $\overline{V}^*$ because the projection $p$ is both open and closed. In particular, $\dot{V}^*$ is its boundary in $E^*$. Let $\partial: V^* \to (0, \infty)$ be half the distance of a point from the boundary of $V^*$ measured in some metric $\delta$ on $E^*$, i.e. $\partial(x^*) = \delta(x^*, \dot{V}^*)/2$. Choosing a smooth map $f_{1/2}{}^*$ close enough to $f^*$, it will be homotopic to $f^*$ and we can arrange that this homotopy moves less than $\partial$ away from $f^*$ ([Bröcker-Jänich], 12.9 and 14.8). Using R. Thom's transversality theorem, we find, arbitrarily close to the section $x^* \mapsto (x^*, f_{1/2}{}^*(x^*))$ of the projection $V^* \times E^* \to V^*$, a section $\varphi_1{}^*(x^*) = (x^*, f_1{}^*(x^*))$ which comes transversally to the diagonal in $V^* \times E^*$. Then $f_1{}^*$ stays arbitrarily close to $f_{1/2}{}^*$. In particular, $f_{1/2}{}^*$ and $f_1{}^*$ are homotopic and again, we may assume that the homotopy moves points by less than $\partial$.

Composing these two homotopies, we get a homotopy $f_t^*$ from $f^*$ to $f_1{}^*$ with the property $\delta(f_t^*(x^*), f^*(x^*)) < \delta(x^*, \dot{V}^*)$ for all $(x^*, t) \in V^* \times [0, 1]$. In particular, $\delta(f_t^*(x^*), f^*(x^*))$ tends to 0 as $x^*$ approaches the boundary of $V^*$. Hence, at each instant $t$, we can use $f^*$ to extend $f_t^*$ continuously to $\dot{V}^*$: On $\overline{V}^*$, we thus get a homotopy $f_t^*$ of $f^*$ relative to $\dot{V}^*$ such that, in $V^*$, the fixed points of $f_1{}^*$ form a discrete set. For, $\mathrm{Fix}(f_1{}^* | V^*)$ is a submanifold of $V^*$ of codimension $\dim(V^*) = \dim(E^*)$ being the preimage of the diagonal in $V^* \times E^*$ under $\varphi_1{}^*$. The Steenrod CHP-theorem now provides us with a $G$-homotopy $f_t$ of $f$ over $f_t^*$ which is stationary on $\dot{V}$. Its fixed point set is a compact subspace of $V \times [0, 1]$ because $f$ is fixed point free on the boundary of $V$. Therefore, the restrictions of $f$ and $f_1$ to $V$ have one and the same index. But the fixed points of $f_1$ are spread over only a finite number of orbits in $V$ since $(\mathrm{Fix}(f_1))^* \subset V^*$ is a compact subspace of the fixed point set of

$f_1{}^*$ whose portion in $V^*$ is discrete: So, $\mathrm{Fix}(f_1)$ consists of a finite number of full fibres of the bundle $p: E \to E^*$.

If we now separate the points in $(\mathrm{Fix}(f_1))^*$ by sufficiently small open neighbourhoods $X^* \subset Y^*$ with $f_1{}^*(X^*) \subset Y^*$, then, in a $G$-neighbourhood of its fixed points set, $f_1$ splits into a finite number of topological summands $f_X: G \times X^* \to G \times Y^*$ in such a way that $f_X{}^*: X^* \to Y^*$ has exactly one fixed point. As a $G$-map, $f_X$ maps the fibres $G$ by right translation, and over the only fixed point of $f_X{}^*$, $f_X$ has to be the identity.

Hence, the composite $g \circ f_X$ of $f_X$ with the right multiplication by $g^{-1}$ in the fibre $G$ is fixed point free for every non-trivial $g \in G$. On the other hand, when $g$ belongs to the unit component of $G$, then $g \circ f_X$ is $G$-homotopic to $f_X$ via a compactly fixed $G$-homotopy. Therefore, all the indices $I_G(f_X)$ and hence their sum $I_G(f)$ vanish unless $G$ is finite.

For finite $G$ now, the translation factor $g \in G$ of $f_X$ in some fibre $G$ over $X^*$ will not depend on the fibre considered when we have chosen $X^*$ to be connected. Whence, $f_X$ must then be the identity in each fibre over $X^*$, i.e. $f_X = \mathrm{id}_G \times f_X{}^*$. Let $e: G \to \{e\}$ denote the constant homomorphism. ADD and MUL yield $I_G(f) = \sum_X \chi_G(G) \cdot I_G(e^*(f_X))$ and as $\pi_{\underline{G}}$ is an equivariant cohomology theory, NAT implies $I_G \circ e^* = e^* \circ I$ where $I \in \mathbb{Z}$ is the Hopf index. Altogether, we have

$$I_G(f) = I(\sum_X f_X{}^*) \cdot \chi_G(G)$$

as asserted, for $e^*(1)$ is $1_{\pi_G}$. □

**5.9** COMMENT. *The integral factor $n(f) \in \mathbb{Z}$ counts the orbits in* $\mathrm{Fix}(f)$ *rather than the fixed points of $f/G$.* $\mathrm{Fix}(f/G)$, the set of $f$-invariant orbits, is the orbit space of the union of all $\mathrm{Fix}(gf)$ taken over $g \in G$. In particular, $\sum_{g \in G} I(gf)$ equals $|G| \cdot I(f/G)$ if $G$ is finite - provided of course, $f/G$ is compactly fixed. We will see in the next section that this holds true even if $G$ does not act freely on $E$.

**5.10 Corollary.** *Let $f: V \to E$ be a $G$-fixed point situation in a $G$-ENR with a single orbit type $(G/H)$. Then the $h_G$-index of $f$ vanishes unless the $G$-automorphism group $W(H) = N(H)/H$ of $G/H$ is finite. In that case, $|W(H)|$ divides the Hopf index of $f^H$ and we have*

$$I_{h_G}(f) = I(f^H)/|W(H)| \cdot \chi_{h_G}(G/H).$$

PROOF. Again, it suffices to prove the statement in $G$-cohomotopy theory $\pi_G$. This time, we exploit that $E$ is the fibre bundle with fibre $E^H$ associated to the $N(H)$-principal bundle $G \to G/N(H)$ (I, 2.11). So, any $G$-fixed point situation $f$ in $E$ has the form $\mathrm{id}_G \times_{N(H)} \omega^*(f^H)$ and $f^H$ is a free $W(H)$-fixed point situation. $\omega$ denotes the projection $N(H) \to W(H)$.
Corollary 4.22 together with NAT yields $I_G(f) = \mathrm{ind}\big(\omega^*(I_{W(H)}(f^H))\big)$ and according to Lemma 5.8, the $W(H)$-index of $f^H$ vanishes unless $W(H)$ is finite in which case it equals $I(f^H)/|W(H)| \cdot \chi_{W(H)}(W(H))$. □

**5.11** Combining the preceding result with 5.4 and 5.2, we obtain the following sum formula for equivariant fixed point indices, thereby generalizing Theorem 5.3.13 in [tom Dieck 2].

**Theorem.** *Let $f$ be a $G$-fixed point situation over a point. The $h_G$-index of $f$ decomposes into an integral linear combination of the $h_G$-characteristics of the orbit types around* Fix($f$):

$$I_{h_G}(f) = \sum_{(H)} n_H(f) \cdot \chi_{h_G}(G/H).$$

*$\chi_{h_G}$ vanishes on all orbit types $(G/H)$ whose $G$-automorphism group $W(H) = N(H)/H$ is not finite. Otherwise, the Hopf index $I(f^H) - I(f^{\underline{H}})$ is divisible by the order of $W(H)$ and we have*

$$n_H(f) = \big(I(f^H) - I(f^{\underline{H}})\big) \,/\, |W(H)|.$$

*Further, if the Euler-Poincaré characteristic of $G/H$ is non-zero, i.e. if $H \leq G$ is a subgroup of maximal rank, then it divides the Hopf index $I(f^{(H)}) - I(f^{\underline{(H)}})$ yielding anew $n_H(f)$:*

$$n_H(f) = \big(I(f^{(H)}) - I(f^{\underline{(H)}})\big) \,/\, \chi(G/H) \;\; \textit{if} \;\; \mathrm{rank}(H) = \mathrm{rank}(G).$$

Observe that for simplicity, we have preferred to write $n_H$ rather than $n_{(H)}$. □

**5.12** Let us inspect our sum formula in ordinary singular cohomology $H^*$: We only find terms with subgroups $H \leq G$ of maximal rank, for otherwise, the Euler-Poincaré characteristic $\chi(G/H)$ is zero. According to a result of [Samelson-Hopf], $\chi(G/H)$ equals $|N(T)\,/\,(N(T) \cap H)|$ if $T \leq G$ is a maximal torus contained in $H$.

Since in a small neighbourhood of Fix($f$), all isotropy groups are subconjugate to isotropy groups living on Fix($f$) (I, 2.11), we deduce

**Corollary.** *Let $f$ be a $G$-fixed point situation over a point. For every torus $S \le G$, we have*

$$I(f) = I(f^S).$$

*In particular, the Hopf index $I(f)$ of $f$ vanishes if none of the isotropy subgroups of $G$ on* Fix($f$) *is of maximal rank. Otherwise, $I(f)$ is divisible by the greatest common divisor $\tau_G(f)$ of the Euler-Poincaré characteristics of the orbits on* Fix($f$).
*If $G$ is a finite group and if $p$ is a prime dividing its order, then modulo $p$, the decomposition of $I(f) - I(f^G)$ only contains terms belonging to such proper subgroups $H \le G$ whose $p$-Sylow subgroups are Sylow subgroups in $G$. For $G$ a finite $p$-group, we thus get*

$$I(f) \equiv I(f^G) \bmod p.$$

□

**5.13** Interpreted in stable $G$-cohomotopy, the sum formula 5.11 purports that every $G$-fixed point situation over a point is equivalent to an integral linear combination of identities on orbits of $G$: $[f] = \sum n_H(f) \cdot [\mathrm{id}_{G/H}]$. Actually, there do only contribute orbit types of finite $G$-automorphism group $W(H)$.
It is obvious now that $f$ is equivalent to the identity on a suitable compact $G$-ENR: For a positive coefficient, take $n_H(f)$ copies of $G/H$ and for a negative one, $(-n_H(f))$ copies of $G/H \times X$ where $X$ is any compact ENR of Euler-Poincaré characteristic $(-1)$. A simple candidate for $X$ is the figure-eight $S^1 \vee S^1$, a more complicated one the connected sum of three projective planes. The disjoint union of all these spaces is a compact $G$-ENR $E$ with the property $\chi(E^H) - \chi(E^{\underline{H}}) = n_H(f) \cdot \chi(G/H^H)$. Since $G/H^H$ equals $W(H)$, Theorem 5.11 shows $\chi_G(E) = I_G(f)$, i.e. $[1_E] = [f]$ in $\mathrm{Fix}_G(\mathrm{pt})$.

**Corollary.** *Over a point, any $G$-fixed point situation is equivalent to the identity on a compact* $G$-ENR. □

Further applications of the sum formula will be found in the next section. Concerning the calculation of the coefficients in the sum formula, we note finally:

5.14 REMARK. Let $f$ be a $G$-fixed point situation in a *compact* $G$-ENR $E$. According to Corollary 5.3, $\left(I_{h_G}(f^{(H)}) - I_{h_G}(f^{\underline{(H)}}) + 1_{h_G}\right)$ is the $h_G$-index of the $G$-fixed point situation induced by $f$ in the $G$-ENR $E^{(H)}/E^{\underline{(H)}}$. Since $E$ is compact, this is the *one-point compactification* $E^c_{(H)}$ of the open $G$-subspace $E_{(H)} \subset E^{(H)}$. Denoting the induced map by $f^c_{(H)}$, we have

$$I_{h_G}(f^{(H)}) - I_{h_G}(f^{\underline{(H)}}) = I_{h_G}(f^c_{(H)}) - 1_{h_G}.$$

*Concerning the coefficients in the sum formula, we get in particular*

$$I(f^H) - I(f^{\underline{H}}) = I(f^c_H) - 1.$$

*When $f$ is defined as a global map, the latter is the reduced Lefschetz number $\tilde{L}(f^c_H)$ of $f^c_H$ with coefficients in some field $F$:*

$$I(f^H) - I(f^{\underline{H}}) = \sum (-1)^i \operatorname{trace}\left(f^* : \tilde{H}^i(E^c_H;F) \to \tilde{H}^i(E^c_H;F)\right)$$

On ENRs, singular cohomology coincides with Čech-cohomology and for any locally compact space $X$, $\check{\tilde{H}}^*(X^c)$ equals $\check{H}_c^*(X)$ ([Spanier], 6.9.5 and 6.7.12). Hence, *in case $f$ maps $E_H$ to itself, $I(f^H) - I(f^{\underline{H}})$ is the unreduced Lefschetz number $L_c(f_H)$ of $f_H\colon E_H \to E_H$* in singular or Čech-cohomolgy *with compact support.* □

5.15 Consider for instance the identity on $E$: $\chi(E^H) - \chi(E^{\underline{H}})$ is the Euler-Poincaré characteristic $\chi_c(E_H)$ with compact support. The latter equals $\chi_c(E_H/N(H)) \cdot \chi(W(H))$ ([tom Dieck 2], 5.2.10; cf Example 6.16) and $\chi_c(E_H/N(H))$ is again $\chi(E^H/N(H)) - \chi(E^{\underline{H}}/N(H))$.
All these equations hold as well with $H$ and $N(H)$ replaced by $(H)$ and $G$. Observing $E_H/N(H) = E_{(H)}/G$, we have

**Corollary.** *Let $E$ be a compact $G$-*ENR. *The sum coefficients of its $h_G$-characteristic read*

$$n_H(E) = \chi_c(E_H) \;/\; |W(H)| = \chi_c(E_{(H)}/G)$$

*or likewise*

$$\chi\left(E^H/N(H)\right) - \chi\left(E^{\underline{H}}/N(H)\right) = \chi\left(E^{(H)}/G\right) - \chi\left(E^{\underline{(H)}}/G\right).$$

*For a subgroup $H \leq G$ of maximal rank, we have further*

$$n_H(E) = \chi_c(E_{(H)}) \,/\, \chi(G/H).$$

*Thus, the Euler-Poincaré characteristics of E and E/G respectively decompose into the following sums over the orbit types on E (cf II, 6.9):*

$$\chi(E) = \sum_{(H)} \chi_c(E_{(H)})$$
$$\chi(E/G) = \sum_{(H)} \chi_c(E_{(H)}/G).$$

*To $\chi(E)$, actually, there do only contribute orbit types $(G/H)$ of non-trivial Euler-Poincaré characteristic.* □

**5.16 Corollary.** *For a compact G-ENR E with a finite group G, we get the decompositions*

$$\chi(E) = \sum_H \chi(E^H) - \chi(E^{\underline{H}}) = \sum_H \chi_c(E_H)$$
$$\chi(E/G)\,|G| = \sum_H \chi_c(E_H/N(H))\,|N(H)| = \sum_H \chi_c(E_H)\,|H|$$

*where H runs through the set of all subgroups of G.*

Proof. This follows at once from $\chi_c(E_H)\,|G/H| = \chi_c(E_{(H)})\,|W(H)|$ (5.10) since the number of subgroups of $G$ conjugate to $H$ is $|G/N(H)|$. □

## 6. Relations between Hopf Indices

**6.1** We list some applications of the sum formula 5.11 in ordinary singular cohomology. As in Section 5, $G$ has to be a *compact Lie group.*

If $X$ is a $G$-space, we write $X^g$ for the fixed point set of $g \in G$. That is, in the superscript, we let $g$ stand for the closed subgroup $\langle g \rangle \leq G$ generated by $g$.

**6.2 Theorem.** *Let $f$ be a $G$-fixed point situation over a point with $G$ a finite group. Then the sum of Hopf indices $I(f^g)$, taken over all $g \in G$, is divisible by the order of $G$:*

$$\sum_{g \in G} I(f^g)/|G| = \sum_{(H)} n_H(f) \in \mathbb{Z}.$$

*$(H)$ - or rather $(G/H)$ - runs through the set of orbit types around* $\mathrm{Fix}(f)$ *and $n_H(f)$ is given by*

$$n_H(f) = \left(I(f^H) - I(f^{\underline{H}})\right) / |W(H)| = \left(I(f^{(H)}) - I(f^{\underline{(H)}})\right) / |G/H|.$$

*For the identity on a compact $G$-ENR $E$, we get in particular*

$$\sum_{g \in G} \chi(E^g) = \chi(E/G)\,|G|.$$

PROOF. According to Theorem 5.13, $f$ is equivalent to the sum $\bigoplus_{(H)} n_H(f) \cdot [\mathrm{id}_{G/H}]$ taken over all orbit types of $G$. Hence, $I(f^g)$ equals $\sum_{(H)} n_H(f)|G/H^g|$ for every $g \in G$. Since a coset $\gamma \in G/H$ is fixed under $g \in G$ if and only if $g$ belongs to the isotropy group $G_\gamma$ at $\gamma$, $\sum_{g \in G}|G/H^g|$ equals $\sum_{\gamma \in G/H}|G_\gamma| = |H|\,|G/H|$, which proves the first statement. For $f = \mathrm{id}_E$, the sum of all $n_H(f) = n_H(E)$ just yields $\chi(E/G)$ by Corollary 5.15. □

**6.3** Let $G$ be a compact Lie group again. If $f$ is a $G$-fixed point situation and $H \leq G$ is a subgroup of finite index in its normalizer, then the sum of the Hopf indices $I((f^H)^w)$, taken over $w \in W(H)$, is divisible by the order of $W(H)$. We rearrange the sum:
All generators $w$ of some cyclic subgroup $K/H \leq W(H)$, $K \leq N(H)$, give rise to one and the same map $(f^H)^w = f^K$. And due to COM, $I(f^K)$ depends only on the conjugacy class of $K$, for $f^{gKg^{-1}}$ agrees with $gf^Kg^{-1}$. Since the number of subgroups of $N(H)$ conjugate to $K$ in $N(H)$ is $|N(H)\,/\,(N(H) \cap N(K))|$, we have the following result:

**Corollary.** *Let $f$ be a $G$-fixed point situation over a point with $G$ a compact Lie group. Then, for every subgroup $H \leq G$ of finite index in its normalizer, the Hopf indices of $f$ satisfy the congruence*

$$\sum_{(K)} |\mathrm{gen}(K/H)|\ |N(H)/(N(H)\cap N(K))|\cdot I(f^K) \equiv 0 \mod |W(H)|$$

*where the sum is taken over those conjugacy classes in G which have a representative K containing H as a normal subgroup and such that K/H is cyclic.* □

**6.4** Example. For the sphere $S$ in a $G$-module $M$, we have $\chi(S^g) = 1 - (-1)^{\dim(M^g)}$. Hence, if $G$ is finite, *the dimensions of the linear subspaces $M^g$ either must all be of one and the same parity or there have to be among them as many of even dimension as there are of odd dimension.* In particular, for any automorphism $g$ on $M$ of *odd* order, we have $\dim(M^g) \equiv \dim(M) \mod 2$.

Further, $G = \mathbb{Z}/2$ *is the only non-trivial finite group to act quasi-freely by linear transformations on a vector space of odd dimension. More generally, among all non-trivial groups, only this one can act freely on a sphere of even dimension:*

In case $G$ is a compact Lie group, this follows at once from $\chi(S) = \chi(S/G)\,\chi(G)$. Otherwise, every non-trivial $g \in G$ has the mapping degree $-1$ because $I(g) = 1 + \deg(g)$ is zero. Since the composite of two orientation reversing maps is orientation preserving, we must have $g^{-1}g' = e$ for any two non-trivial elements $g$ and $g'$ in $G$.

**6.5** From the relation in 6.2, we can easily derive some Borsuk-Ulam type consequences:

**Proposition.** *Let $f\colon V \to E$ be an odd fixed point situation in a symmetrical* ENR $E \subset \mathbb{R}^n$ - i.e. if $x \in \mathbb{R}^n$ is in $V$ or in $E$ respectively, then so is $-x$, and $f(-x)$ equals $-f(x)$.
*Then the Hopf index $I(f)$ is odd if and only if $0 \in V$. In particular, if 0 is not in $V$, then $f$ does not admit a compactly fixed contraction to any point in $V$.*

Proof. Let $G := \mathbb{Z}/2$ act on $\mathbb{R}^n$ via multiplication by $-1$. Then $E$ is a $G$-ENR since $E$ and $E^G$ are ENRs, and $f$ is a $G$-map. Hence, by 6.2, $I(f)$ is congruent to $I(f^G)$ modulo 2 and $I(f^G)$ is 1 exactly in the case $0 \in V$. □

**6.6** This readily implies the classical antipodal theorem of Borsuk-Ulam:

**Corollary.** *Any odd transformation of a sphere is surjective. In particular, $S^n$ admits an odd mapping to $S^m$ if and only if $n \leq m$.*
*Hence, given any map $f\colon S^n \to \mathbb{R}^n$, there exists some $x \in S^n$ such that $f(-x) = f(x)$.* □

**6.7** Proposition 6.5 generalizes to *quasi-free* group actions as follows:

**Proposition.** *Let $E$ be a space whereon $G$ acts freely with the exception of one point $c$,* which consequently is fixed under $G$. According to the Jaworowski criterion (II, 4.6), *$E$ is a $G$-*ENR *if and only if it is an* ENR.
*If $G$ is finite, then modulo $|G|$, the Hopf index $I(f)$ of any $G$-fixed point situation in $E$ equals* 1 *if $f$ is defined on $c$, and* 0 *otherwise.*

PROOF. If $f$ is defined on $c$, then $f^g$ is $\mathrm{id}_{\{c\}}$ for all non-trivial $g \in G$, whence $\sum I(f^g) = I(f) + |G| - 1$. Otherwise, we have $I(f^g) = 0$ for $g \neq e$. Alternatively, the statement follows directly from Theorem 5.11. □

**6.8** EXAMPLE. We equip $\mathbb{R}^{2n+2} = \mathbb{C}^{n+1}$ with the following linear $\mathbb{Z}/r$-action: Take any $n$ integers $q_1, q_2, \dots, q_n$ and form the diagonal matrix with entries $\exp(2\pi i/r)$, $\exp(2\pi i q_1/r), \dots,$ $\exp(2\pi i q_n/r)$. If each $q_\nu$ is prime to $r$, then the action of $\mathbb{Z}/r$ on $\mathbb{R}^{2n+2}$ is what we have called quasi-free, and we may apply Proposition 6.7 with $E \subset \mathbb{R}^{2n+2}$ any $\mathbb{Z}/r$-subspace:
If $E$ is an ENR and if $f: V \to E$ is an ordinary fixed point situation which is a $\mathbb{Z}/r$-map, then *the Hopf index $I(f)$ is a multiple of $r$ provided* 0 *is not in $V$*. In that case, $f$ does not admit a compactly fixed contraction to any point in $V$.
The orbit space of $S^{2n+1}$ is known as the *lens space of type* $(r, q_1, \dots, q_n)$. So, *any lens transformation on $S^{2n+1}$ is surjective.*

**6.9** Proposition 6.7 implies at once:

**Corollary.** *Let $M$ be a $G$-module with $G$ finite and non-trivial, and let $S(M)$ denote the sphere in $M$.*
*A $G$-transformation on $S(M)$ is never null-homotopic, hence always surjective. In particular, there does not exist a $G$-mapping from $S(M)$ to the sphere in any proper $G$-submodule $N \subset M$. Whence, any $G$-map $S(M) \to N$ hits* 0 *and consequently, for any map $f: S(M) \to N$, there exists some point $x \in S(M)$ such that $\sum_{g \in G} g^{-1} f(x)\, g = 0$.* □

**6.10** EXAMPLE. ([Gottlieb]). *Set $g := \exp(2\pi i/r)$. For any map $f: S(\mathbb{C}^{n+1}) = S^{2n+1} \to \mathbb{C}^{n+1}$ there exists a point $x \in S^{2n+1}$ such that*

$$\sum_{\nu=1}^{r} g^{-\nu} f(g^{\nu} x) = 0 \quad \text{or} \quad \sum_{\nu=1}^{r-1} g^{-\nu}\big(f(g^{\nu} x) - f(x)\big) = 0.$$

Thus, $f$ maps the orbit $\{g^v x\}$ of $x$ to a *complex hyperplane in* $\mathbb{C}^n$ *of dimension* $r-2$.
If $f$ is real valued, i.e. $f: S^{2n+1} \to \mathbb{R}^n$, then the orbit of $x$ gets mapped to an $(r-3)$-dimensional *real hyperplane of* $\mathbb{R}^n$. For, the real and the imaginary part of the above complex relation yield two linearly independent real relations between the vectors $f(g^v x) - f(x)$. □

**6.11** The following amplification of the antipodal theorem in 6.9 is an immediate consequence of Corollary 5.12.

**Proposition.** *Let $M$ be an $m$-dimensional $G$-module with $G$ a compact Lie group and let $f$ be a $G$-transformation on the sphere $S(M)$ in $M$. For the mapping degree of $f$, we get*

$$\deg(f) = (-1)^m$$

*if the Euler-Poincaré characteristics of the orbits of $G$ on* Fix($f$) *vanish all, and*

$$\deg(f) \equiv (-1)^m \mod \tau_G(f)$$

*otherwise, whereby $\tau_G(f)$ denotes their greatest common divisor. In any case, $f$ is not null-homotopic and hence surjective.*
*In particular, $M$ must be of even dimension if the Euler-Poincaré characteristic of each orbit on $S(M)$ vanishes or if their greatest common divisor $\tau_G(M)$ is greater than 2. Finally, we have*

$$\deg(f) \equiv 1 \mod \tau_G(M),$$

for $\tau_G(M)$ divides $\tau_G(f)$. □

**6.12** Let $f$ be any $G$-fixed point situation in a $G$-ENR $E$. We want to investigate the relationship between the Hopf index of $f$ and that of the transformation $f/G$ induced on orbit spaces.
First, $E/G$ is an ENR ([tom Dieck 2], 5.2.5), and in case $G$ is finite, $f/G$ is compactly fixed if and only if all the maps $gf$ are so because $\mathrm{Fix}(f/G)$ is $\left(\bigcup_{g\in G} \mathrm{Fix}(gf)\right)/G$. The latter motivates the following assertion:

**Theorem.** *Let $f$ be a partial G-transformation on a G-*ENR *$E$ with $G$ a finite group. If $f/G$ is compactly fixed, then all the maps $gf$, $g \in G$, are fixed point situations in E and $I(f/G)$ is the average of their Hopf indices:*

$$\sum_{g \in G} I(gf) = I(f/G)\,|G|.$$

PROOF. As before, we write $f^*$ for $f/G$. First of all, we may assume that there is only one orbit type on $E$:
As in 5.4, $I(gf)$ splits into the sum $\sum_{(H)} I((gf)^{(H)}) - I((gf)^{\underline{(H)}})$ taken over the orbit types on $E$, for each $E^{(H)}$ is invariant under $gf$. According to Lemma 5.2, the $(H)^{\text{th}}$-summand is the index of a fixed point situation $(gf)_{(H)}$ in $E_{(H)}$ which came out as the shrinking of the composite of $gf$ with a suitable $G$-deformation $d_{(H)}$ of $V^{(H)}$: Hence $(gf)_{(H)}$ agrees with $gf_{(H)}$. The map $f_{(H)}{}^*$, induced by the $G$-fixed point situation $f_{(H)}$ in $E_{(H)}$, is a fixed point situation in $E_{(H)}{}^*$ as it is the shrinking of $f^* d_{(H)}{}^*$. By 5.2, its index reads $I((f^{(H)})^*) - I((f^{\underline{(H)}})^*)$ and as in 5.4, the sum of these indices taken over the orbit types on $E$ amounts to $I(f^*)$. Thus we have to prove the statement for $f = f_{(H)}$ only.
Since $G$ is finite, $E = E_{(H)}$ consists of $|G/N(H)|$ disjoint copies of the closed subspace $E^H$. Hence, $E^H$ is open in $E$ and for each $g \in G$, $I(gf)$ splits into the sum of the indices $I(gf\,|\,\bar{g}V^H)$ taken over representatives $\bar{g}$ of the cosets of $G$ by $N(H)$. On $\bar{g}V^H$ now, $gf$ has the same index as $(\bar{g}^{-1}g\bar{g})f$ on $V^H$. This follows by applying COM to $\bar{g}^{-1}gf\colon \bar{g}V^H \to E$ and $\bar{g}\colon E \to E$. Whence, $\sum_{g \in G} I(gf) = \sum_{g \in G} I(gf^H)|G/N(H)|$. Since $gf^H$ is fixed point free if $g$ is not in $N(H)$, this sum equals $\sum_{n \in N(H)} I(nf^H)|G/N(H)| = \sum_{w \in W(H)} I(wf^H)|G|/|W(H)|$.
Thus, the statement remains to be proved for a free $G$-fixed point situation, for $f^H$ is free on $W(H)$ and $f^H/W(H)$ is the same as $f/G$. We rely on Proof 5.8:
Since $f^*$ is compactly fixed, we may assume that $f^*$ is defined on $\overline{V}^*$ and has no fixed points on the boundary of $\dot{V}^*$. Then $f_t^*|V^*$ is a compactly fixed homotopy starting with $f^*|V^*$ since on $\dot{V}^*$, $f_t^*$ agrees stationarily with $f^*$. On $V$ hence, each $gf$ is connected to $gf_1$ by a compactly fixed homotopy and the fixed point set of $f_1^*|V^*$, which was discrete, is compact and hence finite. Therefore, we may assume that the orbits of $G$ invariant under $f$ are finite in number, say $Gx_i$ for $i \in \underline{r}$.
Then, $\sum_{g \in G} I(gf)$ is the sum of the local indices $I(gf; g'x_i)$ taken over $(g, g', i) \in G \times G \times \underline{r}$. By COM, this sum equals $\sum I(gf; x_i)|G|$, taken over $g$ and $i$. Since $G$ acts freely, there is exactly one map $g_i f$ keeping $x_i$ fixed and by virtue of TOP, the local

index of $g_i f$ at $x_i$ is the same as that of $f^*$ at $x_i^*$. For, as a locally trivial bundle with a discrete fibre, the projection $E \to E^*$ is a local homeomorphism.
Altogether, we get $\sum_{i \in \underline{L}} I(f_i^*; x_i^*) = I(f^*)$ for our sum $\sum_{g \in G} I(gf)/|G|$. □

6.13 REMARK. In the above proof, we could have avoided the reduction to the free case if we had carried out the construction in Proof 5.8 for a $G$-fixed point situation with just one but arbitrary orbit type. This would not have caused any problems apart from a complicated notation, but it was not required there in view of the subsequent Corollary 5.10.

6.14 In general, the above result does not make any sense for $G$-indices. If, however, $G$ is an abelian group, then each $gf$ is a $G$-map. According to Theorem 5.4, its $h_G$-index splits into the sum $\sum_H I_{h_G}((gf)_H)$, taken over the isotropy subgroups of $G$ on $E$, and by Corollary 5.10, we have $I_{h_G}((gf)_H) = I((gf)_H)/|G/H| \cdot \chi_{h_G}(G/H)$. Theorem 6.12 thus implies

**Corollary.** *Let $f$ be a $G$-fixed point situation in a $G$-ENR $E$ such that $f/G$ is compactly fixed and suppose that $G$ is a finite abelian group. With $n_H(f/G) := I(f^H/G) - I(f^{\underline{H}}/G)$ we have*

$$\sum_{g \in G} I_{h_G}(gf) = \sum_H n_H(f/G)|H| \cdot \chi_{h_G}(G/H),$$

$$I(f/G) = \sum_H n_H(f/G)$$

*where the sums are taken over all isotropy subgroups $H \leq G$ on $E$. If all orbits on $E$ are of type $(G/H)$, we get*

$$\sum_{\bar{g} H \in G/H} I_{h_G}(\bar{g} f) = I(f/G) \cdot \chi_{h_G}(G/H).$$

*This holds as well if $G$ is a compact abelian Lie group - i.e. essentially a torus -* since according to Corollary 5.10, all terms vanish if $G/H$ is not finite. □

6.15 Theorem 8.18 in [Dold 3] which describes the behaviour of the fixed point transfer under vertical composition of fixed point situations, yields another relation between the indices of $f$ and $f/G$ provided $E$ is a vertical ENR over $E/G$. As shown in Corollary II, 3.12, there is then only one orbit type on $E$, locally at least. For a single orbit type, indeed, we can derive the result of Dold's theorem ad hoc.

**Corollary.** *Let $f$ be a $G$-fixed point situation in a $G$-ENR $E$ with $G$ a compact Lie group such that $f/G$ is compactly fixed. Suppose there is only one orbit type $(G/H)$ on $E$ and $W(H)$ is finite.*
*Then* $\mathrm{Fix}(f)$ *has a $G$-neighbourhood $U \subset E$ which contains no further $f$-invariant orbits, and we have*

$$I_{h_G}(f) = I(f/G \mid U/G)\cdot\chi_{h_G}(G/H).$$

*Thus we get*

$$I_{h_G}(f) = I(f/G)\cdot\chi_{h_G}(G/H)$$

*in case $f$ keeps fixed, point-by-point, each of its invariant orbits.*

PROOF. Suppose that $U^H \subset E^H$ is a $W(H)$-neighbourhood of $\mathrm{Fix}(f^H)$ without any further $f^H$-invariant orbit. Then $U := G\,U^H \subset E$ is a $G$-neighbourhood of $\mathrm{Fix}(f)$ without further $f$-invariant orbits and $(f/G)|(U/G)$ agrees with $(f^H/W(H))|(U^H/W(H))$. Thus, observing $I_{h_G}(f) = I(f^H)/|W(H)|\cdot\chi_{h_G}(G/H)$ (5.10), we must prove the statement for Hopf indices of free $G$-fixed point situations only.
In the case $(H) = \{e\}$, $G$ is finite by assumption. We have seen in Proof 5.8 that then, around any orbit $x^* \in (\mathrm{Fix}(f))^*$, $f$ takes the form $\mathrm{id}_G \times f^* : G \times X^* \to G \times Y^*$ because $E \to E^*$ is a locally trivial $G$-principal bundle. Hence, $X^*$ hits $\mathrm{Fix}(f^*)$ in $(\mathrm{Fix}(f))^*$ only. Therefore, $\mathrm{Fix}(f^*)$ is open in $(\mathrm{Fix}(f))^*$, i.e. there exists a $G$-neighbourhood $U$ of $\mathrm{Fix}(f)$ such as claimed. Let $f'$ be $f|U$. Then Proposition 6.12 yields $I(f') = I(f'/G)|G|$ because $gf$ is fixed point free for $g \neq e$ if $G$ acts freely. □

**6.16** EXAMPLE. *For a compact $G$-ENR $E$, the above corollary yields the formula*

$$\chi_c(E_{(H)}) = \chi_c(E_{(H)}/G)\cdot\chi(G/H)$$

used in Proof 5.15, provided the orbit structure of the one-point compactification $E^c_{(H)}$ is *conical at the point* $c$, as is the case if $E$ is a smooth $G$-manifold.
For, $\chi_c(E_{(H)})$ is the Euler-Poincaré characteristic $\chi(E^c_{(H)}) - 1$ (5.14) which, according to Lemma 5.2, can be calculated as the Hopf index of $d_1|E_{(H)}$ with $d_t$ a suitable $G$-deformation of $E^c_{(H)}$. Analogously, $\chi_c(E_{(H)}{}^*) = I(d_1{}^*|E_{(H)}{}^*)$. Now, if $c$ has a cone-like $G$-neighbourhood $S \times [0,\infty)/S \times \{0\}$ where $G$ acts trivially on $[0,\infty)$ and $c$ corresponds to the cone vertex,

it is obvious that $d_t$ can be chosen so that $d_1$ keeps fixed, point-by-point, each orbit which gets mapped to itself. Corollary 6.15 then implies $I(d_1 | E_{(H)}) = I(d_1{}^* | E_{(H)}{}^*) \cdot \chi(G/H)$ as asserted. □

**6.17** For a compact $G$-ENR $E$ with $G$ a finite group, the sums $\sum_{g \in G} I(g)$ and $\sum_{g \in G} \chi(E^g)$ agree due to the Theorems 6.2 and 6.12. In fact, $I(g)$ equals $\chi(E^g)$ as shown in [tom Dieck] or [Brown]. We are now ready to generalize this result to elements of infinite order.

**Theorem.** *Let $E$ be a compact $G$-ENR with $G$ a compact Lie group. Then, for every $g \in G$, the Hopf index of the multiplication with $g$ on $E$ equals the Euler-Poincaré characteristic of its fixed point set $E^g$:*

$$I(g) = \chi(E^g).$$

PROOF. For finite groups $G$, we proceed by induction on their order: For $G = \{e\}$, the statement is true and else, we know $\sum_{g \in G} I(g) = \sum_{g \in G} \chi(E^g)$. If the order of $g$ is less than $|G|$, the inductive hypothesis says $I(g) = \chi(E^g)$, for $E$ is a $\langle g \rangle$-ENR. So we only have to regard a cyclic group $G$ and generators $g$ therein. Since $E^g$ then equals $E^G$, is suffices to show that $I(g)$ takes one and the same value for all generators $g$ of $G$.
According to the Lefschetz-Hopf trace formula, $I(g)$ is the character of the virtual representation $\sum(-1)^i H^i(E; \mathbb{C})$ of $G$ over $\mathbb{C}$. Since $G$ is cyclic, of order $r$ say, $G$ has exactly $r$ irreducible representations over $\mathbb{C}$, all of dimension one, namely $\theta_v(g) = \exp(2\pi i v/r)$, $v \in \underline{r}$, where $g$ is a preferred generator of $G$. The value of any virtual representation $\theta$ of $G$ on any other generator is obtained from $\theta(g)$ by substituting for $\exp(2\pi i/r)$ a suitable $r^{\text{th}}$-root of unity, in other words, by subjecting $\theta(g)$ to a Galois automorphism of $\mathbb{Q}(\exp(2\pi i/r))$ over $\mathbb{Q}$. But the Hopf index, being an integer, remains unchanged under Galois automorphisms over $\mathbb{Q}$.
In the general case, too, we regard $E$ as a $\langle g \rangle$-ENR, that is we assume $G = \langle g \rangle$. Then multiplication by $g$ is a $G$-transformation of $E$ and according to Theorem 5.4, its index splits into the sum $\sum (I(g^H) - I(g^{\underline{H}})) = \sum (I(g^c_H) - 1)$ (5.14). The term corresponding to $H = G$ has the form $\chi(E^G) = \chi(E^g)$ while all other terms vanish:
If $W(H) = G/H$ is not finite, we refer to Corollary 5.10 and otherwise, we use the initial part of the proof: For, according to the Jaworowski criterion (II, 4.6), $E^c_H$ is a compact $G/H$-ENR on which $gH$ acts like $g$. So we already know that $I(g^c_H)$ equals $\chi((E^c_H)^g)$ when $G/H$ is finite. And for $H \leq G$, i.e. $g \notin G$, $(E^c_H)^g$ consists of just one point namely $c$. □

**6.18** The preceding result gives rise to the following question:

PROBLEM (A. Dold). *Given a fixed point situation $f$ such that* Fix($f$) *is an* ENR. *Which is the adequate notion of an "$f$-ENR" to guarantee*

$$I(f) = \chi(\mathrm{Fix}(f)) \ ?$$

**6.19** For illustration, we enclose some folklore results from the theory of compact Lie groups. Let $G$ be a compact connected Lie group and let $T \leq G$ denote a maximal torus.

[Weil]. *All elements of $G$ are conjugate to $T$.*

PROOF. Take some $g \in G$ and consider the multiplication with $g$ on $G/T$. By HTP, its Hopf index is $\chi(G/T) = |W(T)|$ because $G$ is connected. Hence, $g$ must have a fixed point: But $g\bar{g}T = \bar{g}T$ just says $g \in \bar{g}T\bar{g}^{-1}$. □

[Hopf]. *The $k^{\text{th}}$ power map $p_k(g) := g^k$ on $G$ is surjective. Its Hopf index is $(1-k)^r$ when $r$ is the rank of $G$.*

PROOF. $p_k$ is a $G$-transformation on $G$ with regard to conjugation, whence $I(p_k) = I(p_k^T)$ by 5.12. Since $T \leq G$ is a maximal abelian subgroup, $G^T$ equals $T$. On $S^1 \leq T$, $p_k$ has the index $1-k$. Thus, MUL yields $(1-k)^r$ for the index of $p_k$ on $T = (S^1)^r$ and hence on $G$. □

[Gottlieb]. *If $E$ is a connected $G$-*ENR *which admits a $G$-fixed point situation of non-zero Hopf index, then the action of $G$ on $E$ can be lifted to any regular covering of $E$.*

PROOF. We only have to check whether, for some base point $x \in E$, the evaluation $g \mapsto gx$: $G \to E$ induces the trivial homomorphism on fundamental groups (see [Gottlieb]). Now, the inclusion $T \leq G$ induces an epimorphism on fundamental groups and, a suitable base point $x$ chosen, the evaluation is constant on $T$: For, $E^T$ is non-empty if the index $I(f) = I(f^T)$ (5.12) of some $G$-fixed point situation in $E$ is non-trivial. □

# 7. Regular Fixed Points and Local Hopf Indices

**7.1** For linear $G$-fixed point situations, we can deduce the important relation 6.2 via an elementary approach provided $G$ is abelian. And the general case reduces to the linear one by virtue of the equivariant transversality theorem I, 3.2:

**Proposition.** *Any $G$-fixed point situation over a point with $G$ finite is equivalent to a regular one.* To be precise:
*Let $f$ be a $G$-fixed point situation in a $G$-module $M$ with $G$ a compact Lie group and suppose, for each isotropy subgroup $H \leq G$ around* Fix($f$), *the number of subgroups conjugate to $H$ is finite. Then, on some $G$-neighbourhood $V$ of* Fix($f$), *there exists a compactly fixed $G$-homotopy connecting $f$ to a smooth $G$-map $\tilde{f}$ which has only regular fixed points. In other words, $i - \tilde{f}: V \to M$ comes transversally to the origin in $M$ whereby $i$ denotes the inclusion.*

PROOF. Using COM, the first statement follows from the rest. Since $f$ is compactly fixed, we find a bounded open $G$-neighbourhood $V \subset M$ of Fix($f$) such that $f$ is still defined on its boundary $\dot{V}$. And by hypothesis, we may assume that $G/N(H)$ is finite for all isotropy groups $H$ occurring on $V$.
Let $f''$ be a smoothing of $f$ on $\bar{V}$, i.e. a continuous map smooth on $V$ and equal to $f$ on $\dot{V}$. Then $f' := \int_G g f''(g^{-1}x)$ is a $G$-smoothing of $f$ on $\bar{V}$ and $i - f'$ satisfies the assumptions of Corollary I, 3.3: Thus, there exists a continuous $G$-map $\tilde{f}$ on $\bar{V}$ which, on $\dot{V}$, coincides with $f'$ and hence with $f$, which is smooth on $V$, and which has only regular fixed points. Finally, restricted to $V$, the linear homotopy from $f$ to $\tilde{f}$ is a compactly fixed $G$-homotopy since it is fixed point free on $\dot{V}$ and since $\bar{V}$ is compact. □

**7.2 Addendum.** In particular, $I(f) = I(\tilde{f})$ *is the local mapping degree of $i - \tilde{f}$ at* 0, *that is*

$$I(f) = \sum_{x \in \mathrm{Fix}(\tilde{f})} \operatorname{sign} \det\left(\mathrm{id} - T_x \tilde{f}\right)$$

([Dold 4]). For every $H \leq G$, $f^H$ is connected to $\tilde{f}^H$ on $V^H$ through a compactly fixed homotopy and $\tilde{f}^H$ is smooth with derivative $T_x(\tilde{f}^H) = (T_x\tilde{f})^H$. If $x$ is a fixed point of $\tilde{f}$ in $V^H$, then $\mathrm{id} - T_x\tilde{f}$ is an isomorphism. Whence, $\mathrm{id} - T_x(\tilde{f}^H) = (\mathrm{id} - T_x\tilde{f})^H$ is a monomorphism and hence an isomorphism. I.e.: Also $\tilde{f}^H$ has only regular fixed points which implies

$$I(f^H) = \sum_{x \in \mathrm{Fix}(\tilde{f}^H)} \operatorname{sign} \det\left(\mathrm{id} - T_x(\tilde{f}^H)\right)$$

for all $H \le G$. Of course, id denotes the identity on $M$ or on $M^H$ respectively. □

**7.3** As to linear $G$-fixed point situations, we rely on the following lemma.

**Lemma.** *Let $g$ be a normal endomorphism of $\mathbb{R}^n$ which has no real eigenvalues. Then any endomorphism $\varphi$ of $\mathbb{R}^n$ commuting with $g$ has non-negative determinant.*

PROOF. The non-real eigenvalues of $g$ come in conjugate complex pairs $(\lambda, \bar{\lambda})$. When $b \in \mathbb{C}^n$ is a complex eigenvector corresponding to $\lambda$, then $\bar{b}$ is an eigenvector for $\bar{\lambda}$, and $\mathbb{C}^n$ is the sum of the eigenspaces $E_\lambda(g) \oplus E_{\bar{\lambda}}(g)$ because $g$ is a normal map.
Since $\varphi$ commutes with $g$, the eigenspaces $E_\lambda$ and $E_{\bar{\lambda}}$ are invariant under $\varphi$. Therefore, we find a complex basis $B_\lambda$ in $E_\lambda$ with respect to which the matrix of $\varphi$ over $\mathbb{C}$ is triangular. Then $\overline{B_\lambda}$ is a basis in $E_{\bar{\lambda}}$ triangulating $\varphi$ since for $b \in B_\lambda$, $\varphi(\bar{b}) = \overline{\varphi(b)}$ is a linear combination of the basis vectors of $\overline{B_\lambda}$ proceeding $\bar{b}$.
The union of all $B_\lambda \cup \overline{B_\lambda}$ provides a basis of $\mathbb{C}^n$ in which $\varphi$ takes the form of a triangular matrix with pairs of conjugate complex triangular matrices aligned along its diagonal. Hence, the eigenvalues of $\varphi$ come in conjugate complex pairs, which shows $\det(\varphi) \ge 0$. □

**7.4** EXAMPLE. If $g$ is an orthogonal automorphism on $\mathbb{R}^n$ of *odd* order which has no non-trivial fixed points, then neither $+1$ nor $-1$ can occur as an eigenvalue of $g$. Hence the determinant of every endomorphism commuting with $g$ is non-negative.
For example, *if $M$ is a quasi-free (orthogonal) $G$-module (see 6.7) and $G$ is a finite group whose order is not a power of* 2, *then every $G$-endormorphism on $M$ has non-negative determinant.* By the way, this holds already *if the order of $G$ is neither* 1 *nor* 2:
For, when a $G$-endormorphism $\varphi$ on $M$ has negative determinant, then $(\mathrm{id} - \varphi)$ is a $G$-fixed point situation with Hopf index $-1$. So from $M^g = \{0\}$ for $g \neq e$, we get $\sum_{g \in G} I(\mathrm{id} - \varphi^g) = |G| - 2$ and Theorem 6.2 implies $|G| \in \{1, 2\}$.

**7.5 Lemma.** *Let $f$ be a linear $G$-fixed point situation - i.e. a $G$-endormorphism on some $G$-module which does not have the eigenvalue $+1$. Then, for any two elements $g$ and $h$ of $G$ commuting with each other, we have*

$$I(f^{gh}) \,/\, I(f) = I(f^{g}) \cdot I(f^{h}).$$

*In particular, $f^{g^2h}$ has the same Hopf index as $f^h$. The index of $f^{g^v}$, for instance, equals $I(f)$ for $v$ even and $I(f^g)$ else. And for an element $g$ of odd order, we get $I(f^g) = I(f)$.*

PROOF. Clearly, a $G$-endormorphism $f$ on a $G$-module $M$ is compactly fixed exactly if $+1$ is not an eigenvalue of $f$. In that case, for every $g \in G$, $I(f^g)$ is the sign of the determinant of $f^g$, that is $(-1)^{\#}$ where $\#$ is the number of real eigenvalues $t > 1$ of $f^g$ ([Dold 4]).
We may assume that $G$ acts orthogonally on $M$. Then, $\pm 1$ are the only real eigenvalues of $g$ for any $g \in G$. Let $f_g$ denote the fixed point situation induced by $f$ in the eigenspace $E_{-1}(g)$, and ${}^{g}\!f$ that induced on the sum of all eigenspaces of $g$ corresponding to eigenvalues other than $\pm 1$. Observe that $f$ maps an eigenspace of $g$ to itself because $f$ is a $G$-map.
Then we have $I(f) = I(f^g)\, I(f_g)\, I({}^{g}\!f)$ and Lemma 7.3 yields $I({}^{g}\!f) = \text{sign}\det(\text{id} - {}^{g}\!f) = 1$, whence $I(f) = I(f^g)\, I(f_g)$ for all $g \in G$. This in turn implies $I(f_{gh}) = I(f_g)\, I(f_h)$ if $g$ commutes with $h$, which proves the lemma:
For, let $f_{g,h}$ denote the fixed point situation induced by $f$ on $E_{-1}(g) \cap E_{-1}(h)$. Since $f_g$ and $f_h$ are linear $\langle h \rangle$- and $\langle g \rangle$-fixed point situations respectively, we know $I_{gh} = I_{gh}^{h}\, I_{gh,h}$ and $I_g\, I_h = (I_g^h\, I_{g,h})\,(I_h^g\, I_{h,g})$. The latter equals $I_g^h\, I_h^g$ because of $I^2 = 1$ and with $f_{gh}^{h} = f_g^{h}$ and $f_{gh,h} = f_h^{g}$, we have as asserted $I_{gh} = I_g\, I_h$. □

**7.6 Corollary.** *Let $f$ be a $G$-fixed point situation with $G$ a finite abelian group. Then the sum of Hopf indices $\sum_{g \in G} I(f^g)$ from 6.2 vanishes if and only if there exists some $g \in G$ such that $I(f^g) = -I(f)$.*
*In that case, $g$ is of even order and may be chosen so as to generate a direct summand in $G$ whose order is a power of 2. Otherwise, all indices $I(f^g)$ coincide and hence sum up to $I(f)\,|G|$.*

PROOF. If $G$ is cyclic, generated by some $g_0$, then, by Lemma 7.5, the sum $\Sigma$ of all $I(f^g)$ equals $(I(f) + I(f^{g_0}))\,|G|/2$, which is either $I(f)\,|G|$ or 0.
In the general case, $G$ is the product of cyclic subgroups $G_v$, $v \in \underline{r}$, whose orders are powers of primes, and by virtue of Lemma 7.5, $\Sigma$ is the product of the partial sums $\Sigma_v := \sum_{g \in G_v} I(f^g)$ up to the sign $I(f)^{r-1}$. Hence, $\Sigma = 0$ exactly if $\Sigma_v = 0$ for one $v$. In this

case, when $g_v$ is a generator of $G_v$, we know $I(f^{g_v}) = -I(f)$ from the above, and by Lemma 7.5, the order of $G_v$ must be even, whence a power of 2. Otherwise, the above yields $\Sigma_v = I(f)|G_v|$ for each $v$. In that case, $\Sigma$ equals $I(f)|G|$ and with $I(f^g) = \pm 1$, we get $I(f^g) = I(f)$ for all $g \in G$. □

7.7 **Proposition.** *If $f$ is a G-fixed point situation in a smooth G-manifold M (see II, 6.4), then Lemma 7.5 and Corollary 7.6 hold unchanged for the local Hopf index of $f$ at a regular fixed point in $M^G$.*

PROOF. If $f$ is smooth at some fixed point $x \in M^G$ and if $\mathrm{id} - T_x f$ is an isomorphism, then $T_x f$ is a linear $G$-fixed point situation in $T_x M$ which is equivalent to the localization of $f$ at the point $x$. □

7.8 APPLICATION. *For abelian groups, we can now re-derive relation 6.2 which says*

$$\sum_{g\in G} I(f^g) \equiv 0 \mod |G|.$$

PROOF. Using COM, we may regard $f$ as a $G$-fixed point situation in a $G$-module and by Proposition 7.1, we may assume that $f$ is a smooth map having only regular fixed points. Then, the fixed point set of $f$ is a finite $G$-set $F$ and for every $g \in G$, $I(f^g)$ is the sum of local Hopf indices $\sum I(f^g; x)$, taken over $x \in F^g$. Since a fixed point $x$ of $f$ is fixed under $g$ if and only if $g$ belongs to its isotropy group $G_x$, we have $\sum_{g\in G} I(f^g) = \sum_{x\in F}\sum_{h\in G_x} I(f^h; x)$. Proposition 7.7 yields $\sum_{h\in G_x} I(f^h; x) = i_x|G_x|$ with $i_x \in \{0, I(f; x)\}$. The sum over $G_x$ does not change if we replace $x$ by $gx$: For, $G_{gx}$ is $G_x$ and on a suitable neighbourhood $U$ of $x$, $f^h$ has the same Hopf index as $g(f^h|U)g^{-1} = f^h|gU$ for all $h \in G_x$ (COM). Hence, $i_x$ must be constant along the orbit $Q$ of $x$, say $i_x = i_Q$, which implies $\sum_{g\in G} I(f^g) = \sum_{Q\in F/G} i_Q|G|$. □

7.9 Let us resume the discussion subsequent to the sum formula 5.4 on how $f^H$ counts a fixed point of $f$ in $E^G$:

**Corollary.** *Let $f$ be a G-fixed point situation in a smooth G-manifold with G a compact Lie group and let x be a regular fixed point of $f$ whose isotropy group $G_x$ is a finite cyclic group. Then, for every $H \leq G_x$, the local Hopf index of $f^H$ at x is given by*

$$I(f^H; x) \,/\, I(f; x) = \left(I(f^{G_x}; x) \,/\, I(f; x)\right)^{|G_x/H|}.$$

*Hence, in case $H$ does not contain the 2-Sylow subgroup of $G_x$, the Hopf index of $f^H$ at $x$ coincides with that of $f$. Otherwise, it equals $I(f^{G_x};x)$, which holds as well if $G_x$ is any finite abelian group.*
*In particular, we have $I(f^H;x) = I(f^{G_x};x)$ for all $H \le G_x$ if $G_x$ is an abelian group of odd order.*

PROOF. Since for every $H \le G_x$, $f^H$ is smooth at $x$ with derivative $(T_x f)^H$, we may regard $f$ as a linear $G_x$-fixed point situation. That is, we may take $G$ to be $G_x$.
First, let $G$ be cyclic. In the case $I(f^G) = I(f)$, Lemma 7.5 yields $I(f^H) = I(f)$ for all $H \le G$ as claimed. Otherwise, $I(f^G)$ equals $-I(f)$. Then $G$ must be of even order and for a generator $g$ of $G$, we have $I(f^{g^\nu}) = (-1)^\nu I(f)$, again from Lemma 7.5. But $\nu$ is of the same parity as the order $|G/\langle g^\nu\rangle| = \gcd(\nu, |G|)$ since $|G|$ is even.
For $G$ abelian, we prove the statement by induction on the order of $G/H$. Since $f^H$ is a $G/H$-fixed point situation with $(f^H)^{G/H} = f^G$, we already know $I(f^H) = I(f^G)$ when $|G/H|$ is prime. For then, $G/H$ is a cyclic group of odd order. Otherwise, we find in $G$ a proper subgroup $K > H$ since for every prime $p$ dividing $|G/H|$, there exists in $G/H$ an element of order $p$. By induction hypothesis, we have $I(f^H) = I(f^K)$ as well as $I(f^K) = I(f^G)$. □

7.10 REMARK. As illustrated by Example 5.6, *the Hopf index of $f$ at a regular fixed point $x$ with finite cyclic isotropy group $G_x$ is not determined by $I(f^{G_x};x)$ if $|G_x|$ is even. Vice versa, $I(f;x)$ does not determine $I(f^{G_x};x)$ either:*
$f(x) := 2x$ is a $\mathbb{Z}/2r$-fixed point situation in $\mathbb{R}^{2rn}$ equipped with the cyclic permutation $s$ of the components in each factor $\mathbb{R}^{2r}$. Its indices are $I(f) = I(f^{s^{2\nu}}) = 1$ and $I(f^s) = I(f^{s^{2\nu+1}}) = (-1)^n$.

At some point $x$ with *finite isotropy group* $G_x$, $f^H$ has the same Hopf index as $f^{S_2(H)}$ for every $H \le G_x$ where $S_2(H)$ denotes the 2-Sylow subgroup of $H$. Therefore, studying $I(f^H;x)$ for arbitrary subgroups $H \le G_x$, we may confine ourselves to (linear) $S$-fixed point situations $f$ with *$S$ a finite abelian 2-group.*
Then, if $S$ is cyclic, Corollary 7.9 yields $I(f^H) = I(f)$ for all proper subgroups $H \le S$ whereas *in general, $I(f^H)$ is not determined even by the indices of $f$ and $f^S$ together:*
On $\mathbb{R}^n \times \mathbb{R}^n$ equipped with the action $s_1(x,y) := (-x, y)$, $s_2(x,y) := (x, -y)$ of the group $S := \mathbb{Z}/2 \times \mathbb{Z}/2$, multiplication by 2 is an $S$-fixed point situation which has the Hopf indices $I(f) = 1 = I(f^S)$, $I(f^{s_1}) = (-1)^n = I(f^{s_2})$, and $I(f^{s_1 s_2}) = 1$. □

# 8. Relations on the Equivariant Fixed Point Ring

**8.1** Throughout the section, $G$ is a *compact Lie group.* By $\Phi(G)$, we denote the set of conjugacy classes of those subgroups of $G$ which have finite index in their normalizer.
We want to describe the coefficient ring $Fix_G(\mathrm{pt})$ of equivariant fixed point theory, abbreviated $F(G)$, as a subring of $\mathbb{Z}^{\Phi(G)}$. For that, we first reformulate some results of Section 5 in terms of the Burnside ring of $G$.

**8.2 Proposition and Definition.** *In* $Fix_G(B)$, *the equivalence classes of the identities on proper* $G$-ENR$_B$s *form a subring.* Indeed, we have seen in Proof 4.2 that $-1$ has as representative the projection $B \times (S^1 \vee S^1) \to B$
This ring is called the *Burnside ring* $Burn_G(B)$ *of* $G$ *over* $B$. We denote its elements by $[p]$ rather than $[\mathrm{id}_p]$. Over a point, we write $A(G)$ for the *Burnside ring*, and $F(G)$ for the *fixed point ring of* $G$.
Since, up to equivalence, $G$-fixed point situations are determined by their index in stable $G$-cohomotopy (Theorem 4.3), $Burn_G(B)$ *arises from the set of proper* $G$-ENR$_B$s *by identifying spaces of the same* $\pi_G$*-characteristic.* □

**8.3** REMARK. From Corollary 4.17, we know that $F(G)$ is isomorphic to $Fix(G)$, the non-equivariant fixed point ring over $G$. And we will see in a moment that $A(G)$ is isomorphic to the classical Burnside ring which justifies our notation.

**8.4 Theorem.** *Corollary 5.13 says*

$$A(G) = F(G).$$

*For any* $H \leq G$, *this implies*

$$Burn_G(G/H) = Fix_G(G/H),$$

for the isomorphism $[f] \mapsto [\mathrm{id}_G \times_H f]\colon F(H) = Fix_H(\mathrm{pt}) \to Fix_G(G/H)$ from Proposition 4.15 maps $A(H) = Burn_H(\mathrm{pt})$ onto $Burn_G(G/H)$. □

**8.5** PROBLEM. Over a *non-trivial base B*, however, it is an open problem to identify those stable $G$-cohomotopy classes which are $\pi_G$-characteristics of proper $G$-ENR$_B$s, even in the case $G = \{e\}$.

**8.6** According to Theorem 5.11, the elements of $A(G) = F(G)$ are integral linear combinations of the classes $[G/H] \in A(G)$ with $(H) \in \Phi(G)$. That is, these classes generate $A(G)$ as an abelian group. Even more:

**Proposition.** *$A(G) = F(G)$ is the abelian group free on the classes $[G/H]$ with $(H) \in \Phi(G)$.* As shown in Theorem 5.11, *the $(H)^{\text{th}}$ coordinate of a G-fixed point situation f is given by*

$$n_H(f) = \left(I(f^H) - I(f^{\underline{H}})\right) / \,|W(H)|,$$

*or, if H is of maximal rank in G, by*

$$n_H(f) = \left(I(f^{(H)}) - I(f^{\underline{(H)}})\right) / \,\chi(G/H).$$

Remember that, for simplicity, we use to write $n_H$ rather than $n_{(H)}$.

PROOF. It remains to show that the system of generators $\{[G/H], (H) \in \Phi(G)\}$ is linearly independent. Consider any linear combination $x = \sum_{(H)\in\Phi(G)} n_H \cdot [G/H]$ with integer coefficients.
Since for every $K \in G$, the assignment $f \mapsto I(f^K)$ defines a ring homomorphism $I^K\colon F(G) \to \mathbb{Z}$, we have $I^K(x) = \sum n_H\, \chi(G/H^K)$. Now, if some of the coefficients $n_H$ are non-trivial, we select among their indices some $(K) \in \Phi(G)$ maximal with respect to the total ordering "$\preccurlyeq$" introduced in I, 1.2. Then, $\chi(G/H^K)$ vanishes for every $(H) \neq (K)$ with $n_H \neq 0$, for $\chi(G/H^K) \neq 0$ would imply $(K) \leq (H)$ and hence $(K) \preccurlyeq (H)$. Consequently, $I^K(x) = n_K |\,W(K)|$ is not zero wherefore $x$ can not be trivial. □

**8.7** Actually, we have shown that the ring homomorphism $(I^H)_{H\leq G}\colon F(G) \to \prod \mathbb{Z}$, $[f] \mapsto (I(f^H))$ is injective. And as probe points, we only had to take representatives $H$ of the classes in $\Phi(G)$. By COM, $f^{gHg^{-1}} = g f^H g^{-1}$ has the same Hopf index as $f^H$ for all $g \in G$. Hence, the family $\{I^H, H \leq G\}$ induces a ring homomorphism $I^*$ from $F(G)$ to the set $\mathbb{Z}^{\Phi(G)}$ of maps $\Phi(G) \to \mathbb{Z}$ and we know:

**Proposition.** *$I^*\colon F(G) \to \mathbb{Z}^{\Phi(G)}$, $[f] \mapsto (I(f^H))$ is a monomorphism of unitary rings.* □

**8.8** COMMENT. Therefore, *two G-fixed point situations f and f' have the same $\pi_G$-index if and only if for all $(H) \in \Phi(G)$, the Hopf-indices of $f^H$ and $f'^H$ coincide.* In that case, we have $I(f^H) = I(f'^H)$ and $I(f^{\underline{H}}) = I(f'^{\underline{H}})$ *for all* $H \leq G$.
In particular, the $\pi_G$-characteristics of two compact $G$-ENRs $E$ and $E'$ agree if and only if $E^H$ and $E'^H$ have the same Euler-Poincaré characteristic for all $(H) \in \Phi(G)$, or for all $H \leq G$: So, *$A(G) = F(G)$ is identical with the classical Burnside ring* which is defined as the set of equivalence classes of compact $G$-ENRs with respect to the relation $E \sim E'$ if $\chi(E^H) = \chi(E'^H)$ for all $H \leq G$ (see [tom Dieck 2], 1.2).

**8.9** The fact that already $I^*: F(G) \to \mathbb{Z}^{\Phi(G)}$ is injective, and not just the family of all $I^H$, $H \leq G$, is due to the following result:

**Proposition.** *Let $f$ be a G-fixed point situation over a point. For every $H \leq G$, the Hopf index $I(f^H)$ occurs under the values of $I^*(f)$ on $\Phi(G)$.*
*Indeed, if $T = K/H$ is a maximal torus in $W(H) = N(H)/H$, then $W(K)$ is finite and the Hopf indices of $f^H$ and $f^K = (f^H)^T$ coincide* according to Corollary 5.12.

PROOF. Let $S = L/K$ be a maximal torus in $W(K) = N(K)/K$. We must show that $S$ is trivial. First, $H$ is a normal subgroup of $L$: Take any $l \in L$. There exists a path leading within $L$ from $l$ to $K$ since $L \to L/K$ is a locally trivial fibre bundle over a pathwise connected base. Now, any two sufficiently neighbouring endomorphisms on a compact Lie group are conjugate to each other ([Conner-Floyd], 38.1). Hence, splitting the path from $l$ to $K$ into short segments, we see that on $K$, conjugation with $l$ is an inner automorphism. Since $H$ is normal in $K$, we get $lHl^{-1} = H$.
Thus, $K/H = T$ is a normal subgroup of $L/H$ with quotient $L/K = S$. If we can show that $L/H \leq W(H)$ has to be a torus, we have $K = L$ as claimed since $T$ was a maximal torus in $W(H)$.
Choose $w \in L/H$ so that $S$ is generated by $wT$. $\kappa_w$, the conjugation by $w$, maps $T$ to itself and there, it is homotopic to the identity because $T \lneqq L/H \to S$ is a locally trivial bundle over a pathwise connected base. Since the automorphism group of a torus is discrete, $T$ remains fixed, point-by-point, under $\kappa_w$. Hence, there exists a torus $R \leq L/H$ comprising both $T$ and $w$. And since $S$ is generated by $wT$, $R$ must be the whole of $L/H$. □

**8.10** REMARK. Over $B$, the homomorphism $(I^H)$ may be thought of as the map

$$[f] \mapsto ([f^H]) : \; Fix_G(B) \to \prod_{H \leq G} Fix(B^H).$$

Regarding $I^H$ as a homomorphism $\pi^0_G(B) \to \pi^0(B^H)$, it assigns to some $\varphi: B^+ \wedge S^M \to S^M$ the composite $\varphi^H: (B^H)^+ \wedge S^{M^H} \approx (B^+ \wedge S^M)^H \xrightarrow{\varphi^H} (S^M)^H \approx S^{M^H}$ as follows from NAT.

**8.11** For finite $G$, $F(G) = A(G)$ and $\mathbb{Z}^{\Phi(G)} = \mathbb{Z}\cdot\Phi(G)$ are free abelian groups of the same rank. So, by virtue of $I^*$, $F(G)$ is isomorphic to a subgroup of maximal rank in $\mathbb{Z}\cdot\Phi(G)$. The congruences describing the image of $I^*$ can be read off from the *matrix* $(|G/H^K|)$ *of* $I^*$ *with regard to the bases* $\{[G/H], (H) \in \Phi(G)\}$ *and* $\{(K) \in \Phi(G)\}$. As usual, we let increase the column index $(H)$ from left to right with respect to our total ordering "$\preccurlyeq$" and the row index $(K)$ from top to bottom.

Then the matrix of $I^*$ has no entries below the diagonal since $G/H^K$ is empty unless $(K) \leq (H)$. The diagonal entries are the orders $|W(H)|$. As the $G$-automorphism group of $G/H$, $W(H)$ acts freely on $G/H^K$ wherefore $|W(H)|$ divides the $(H)^{\text{th}}$ column. The reduced coefficients $|G/H^K|/|W(H)|$ are denoted by $G^K_H$.

If $G$ is not finite, we think of $I^*$ as an upper-triangular matrix of infinite size with coefficients $\chi(G/H^K) = |G/H^K|$ for $(H), (K) \in \Phi(G)$. Recall that $G/H^K$ is finite for $(K) \in \Phi(G)$ since for all $H, K \leq G$, it consists of only a finite number of $W(K)$-orbits (I, 2.11). $G^K_H$ is defined like before.

**8.12** We are ready to describe the image of $I^*$:

**Theorem.** *Let G be a compact Lie group. The inverse of the monomorphism* $I^*: F(G) \to \mathbb{Z}^{\Phi(G)}$ *from 8.7 is represented by a three-dimensional matrix of the form* $\mathrm{M} = (m^K_H(\mathscr{H}))$ *where the indices H and K - H for the columns, K for the rows in* M *- are representatives of the classes in* $\Phi(G)$ *and* $\mathscr{H}$ *runs through the finite subsets of* $\Phi(G)$. *In detail,* M *has the following property:*

*A map* $y: \Phi(G) \to \mathbb{Z}$ *belongs to the image of* $I^*$ *if and only if there exists some layer* $\mathrm{M}(\mathscr{H})$ *in* M *such that* $(\mathrm{M}(\mathscr{H})\cdot y)(H) \in \mathbb{Z}$ *is divisible by* $|W(H)|$ *for* $(H) \in \mathscr{H}$ *and vanishes otherwise. In fact,*

$$I^{*-1}(y) = \sum_{(H)} (\mathrm{M}(\mathscr{H})\cdot y)(H)/|W(H)|\cdot[G/H].$$

*Apart from just ones on the diagonal, the only non-trivial entries in the $\mathscr{H}^{\text{th}}$ layer* $\mathrm{M}(\mathscr{H})$ *live in the columns with number $(H) \in \mathscr{H}$. Whence,* $\mathrm{M}(\mathscr{H})$ *represents an endomorphism of $\mathbb{Z}^{\Phi(G)}$, namely*

$$(\mathrm{M}(\mathscr{H})\cdot y)(K) = y(K) + \sum_{(H)} m_H^K(\mathscr{H})\, y(H)$$

*with the sum taken over $(H) \in \mathscr{H} - \{(K)\}$.*

PROOF. We construct M by inverting the matrix of $I^*$: Every $y \in I^*(F(G))$ comes from a free subgroup of $F(G)$ generated by some finite subset $\mathscr{H} \subset \Phi(G)$. Thus, $y$ is an integer linear combination of those columns in $I^* = \left(|G/H^K|\right)$ which are numbered by indices $(H) \in \mathscr{H}$: $y(K) = \sum_{(H)\in\mathscr{H}} n_H |G/H^K|$. Enumerating the elements of $\mathscr{H}$ by size, $(H_1) \succ (H_2) \succ \dots \succ (H_r)$, we read off, starting with $(H_1)$:

$$\begin{array}{ll}
(H_1) \prec (K) & y(K) = 0 \\
(H_1) = (K) & y(K) = n_1 w_1 \\
(H_2) \prec (K) \prec (H_1) & y(K) = n_1 w_1 G_1^K \\
 & 0 = y(K) + \left(-G_1^K\right) y_1 \\
(H_2) = (K) & y(K) = n_2 w_2 + n_1 w_1 G_1^2 \\
 & n_2 w_2 = y(K) + \left(-G_1^2\right) y_1 \\
(H_3) \preccurlyeq (K) \prec (H_2) & y(K) = n_3 w_3 \delta_3^K + n_2 w_2 G_2^K + n_1 w_1 G_1^K \\
 & n_3 w_3 \delta_3^K = y(K) + \left(-G_2^K\right) y_2 + \left(-G_1^K + G_1^2 G_2^K\right) y_2 \\
(H_4) \preccurlyeq (K) \prec (H_3) & y(K) = n_4 w_4 \delta_4^K + n_3 w_3 G_3^K + n_2 w_2 G_2^K + n_1 w_1 G_1^K \\
 & n_4 w_4 \delta_4^K = y(K) + \left(-G_3^K\right) y_3 + \left(-G_2^K + G_2^3 G_3^K\right) y_2 \\
 & \quad + \left(-G_1^K + G_1^2 G_2^K + G_1^3 G_3^K - G_1^2 G_2^3 G_3^K\right) y_1.
\end{array}$$

The index $i$ stands for $(H_i)$, $w_i$ means $|W(H_i)|$, and $\delta_i^K \in \{0, 1\}$ is 1 exactly for $(K) = (H_i)$. The coefficients of the terms $y_i = y(H_i)$ remind us of the *elementary symmetric polynomials*. Writing $G_i 1^K$ instead of $G_i^K$ and $G_i G_j 1^K$ instead of $G_i^j G_j^K$, we may abbreviate the coefficient of $y_i$ in the case $(H_{k+1}) \preccurlyeq (K) \prec (H_k)$ by the *symbol* $(-G_i)\prod_{i<v\le k}(1 - G_v)1^K$. For $i > k+1$, or for $i = k+1$ and $(K) \neq (H_{k+1})$, the symbol also yields the correct coefficient which is

$-G_i^K = 0$. Empty products are 1 by definition. Only in the case $(K) = (H_{k+1})$, the symbol for $i = k+1$ has the value $-G_{k+1}^{k+1} = -1$ rather than 1.
So, we let $m_K^K(\mathscr{H})$ always be 1. Outside the diagonal we set $m_H^K(\mathscr{H}) := 0$ except for the columns with indices $(H) \in \mathscr{H}$. There, we symbolize $m_H^K(\mathscr{H})$ by $(-G_H)\prod(1 - G_{H'})1^K$ where the product is taken over $(H') \in \mathscr{H}$ with $(K) \prec (H') \prec (H)$. Below the diagonal in particular, i.e. for $(K) \succ (H)$, all coefficients vanish. For $(K) \neq (H)$, $-m_H^K(\mathscr{H})$ is given by recursion as

$$-m_H^K(\mathscr{H}) = \sum_{(H')} m_H^{H'}(\mathscr{H})\, G_{H'}^K$$

where the sum is taken over $(H') \in \mathscr{H} - \{(K)\}$, i.e. over $(H')$ with $(K) \prec (H') \preccurlyeq (H)$.
We have to show that the congruences and relations given by M describe the image of $I^*$. Consider any map $y \in \mathbb{Z}^{\Phi(G)}$. If there exists a finite subset $\mathscr{H} \subset \Phi(G)$ such that $(\mathrm{M}(\mathscr{H})\cdot y)(H)$ is divisible by $|W(H)|$ for $(H) \in \mathscr{H}$ and is zero else, then

$$x := \sum_{(H)} (\mathrm{M}(\mathscr{H})\cdot y)(H)/|W(H)| \cdot [G/H]$$

is a well-defined element in $F(G)$. We claim that $I^*(x)$ be $y$. At an arbitrary point $(K) \in \Phi(G)$, $I^*(x)$ takes the value

$$(I^*(x))(K) = \sum_{(H)\in\mathscr{H}} (\mathrm{M}(\mathscr{H})\cdot y)(H)\, G_H^K = \sum_{(H),(H')\in\mathscr{H}} m_{H'}^H(\mathscr{H})\, y(H')\, G_H^K.$$

Splitting off the terms with $(H) = (K)$ and applying the above recursion formula, this takes the form

$$\delta_{K,\mathscr{H}} \sum_{(H')\in\mathscr{H}} m_{H'}^K(\mathscr{H})\, y(H') - \sum_{(H')\in\mathscr{H}-\{(K)\}} m_{H'}^K(\mathscr{H})\, y(H')$$

where $\delta_{K,\mathscr{H}} \in \{0, 1\}$ is 1 exactly if $(K) \in \mathscr{H}$. For $(K) \in \mathscr{H}$, we obtain immediately $(I^*(x))(K) = y(K)$, and otherwise, $(I^*(x))(K) = y(K) - (\mathrm{M}(\mathscr{H})\cdot y)(K)$ agrees with $y(K)$ due to the assumptions made. □

**8.13 Addendum.** *Each layer* $\mathrm{M}(\mathcal{H})$ *in the three-dimensional matrix* $\mathrm{M} = \left(m_H^K(\mathcal{H})\right)$ *is an upper-triangular matrix with ones on the diagonal. Outside the diagonal, the entries are given by the symbolic products*

$$m_H^K(\mathcal{H}) = (-G_H)\prod_{(H')}(1 - G_{H'})\, 1^K$$

*taken over all* $(H') \in \mathcal{H}$ *with* $(K) \prec (H') \prec (H)$. *They are zero unless* $(H) \in \mathcal{H}$ *and* $(H) > (K)$.

Thus, all entries in M arise from the matrix $\left(|G/H^K|\right) = \left(|W(H)|\; G_H^K\right)$ of $I^*$ *by one and the same rule*. For example, if $\mathcal{H}' \subset \mathcal{H}$ is an initial segment in the ordering "$\preccurlyeq$", then the matrix $\mathrm{M}(\mathcal{H}')$ coincides with $\mathrm{M}(\mathcal{H})$ on its support which consists of the diagonal and the columns numbered by $\mathcal{H}'$.

Further, *for* $(K) \neq (H)$, *we are given the recursion formula*

$$-m_H^K(\mathcal{H}) = \sum_{(H')} m_H^{H'}(\mathcal{H})\, G_{H'}^K$$

*where the sum is taken over all* $(H') \in \mathcal{H} - \{(K)\}$, *i.e. over* $(H') \in \mathcal{H}$ *with* $(K) \prec (H') \preccurlyeq (H)$. □

**8.14** For every $(H) \in \Phi(G)$, $|W(H)|^{-1} \cdot I^*(G/H)$ is a map $c_{(H)} : \Phi(G) \to \mathbb{Z}$. And the last part of Proof 8.12 reveals that for any $y \in \mathbb{Z}^{\Phi(G)}$, the map $\sum(\mathrm{M}(\mathcal{H})\cdot y)(H)\cdot c_{(H)}$ agrees with $y$ if $\mathcal{H}$ has the property that $\mathrm{M}(\mathcal{H})\cdot y : \Phi(G) \to \mathbb{Z}$ vanishes on all $(H) \in \mathcal{H}$.

So, for every finite subset $\mathcal{H} \subset \Phi(G)$, we let $C_{\mathcal{H}}$ denote the set of all $c \in \mathbb{Z}^{\Phi(G)}$ for which the map $\mathrm{M}(\mathcal{H})\cdot c : \Phi(G) \to \mathbb{Z}$ vanishes outside $\mathcal{H}$. $C_{\mathcal{H}}$ is a subgroup of $\mathbb{Z}^{\Phi(G)}$ since $\mathrm{M}(\mathcal{H})$ is an endomorphism of $\mathbb{Z}^{\Phi(G)}$. It contains all $c_{(H)}$ with $(H) \in \mathcal{H}$, for $\mathrm{M}(\mathcal{H})\cdot |W(H)|\, c_{(H)}$ equals $\mathrm{M}(\mathcal{H})\cdot I^*(G/H)$.

**Corollary.** *Let* $\mathcal{H}$ *be a finite subset of* $\Phi(G)$. *With the above notation,* $C_{\mathcal{H}}$ *is a free subgroup of* $\mathbb{Z}^{\Phi(G)}$ *with basis* $\{c_{(H)}, (H) \in \mathcal{H}\}$.

*Hence, the image of the subgroup of* $F(G)$ *free on* $\mathcal{H}$ *is a subgroup of maximal rank in* $C_{\mathcal{H}}$ *with index* $\prod_{(H)\in\mathcal{H}} |W(H)|$ *which is described by the congruences*

$$(\mathrm{M}(\mathscr{H})\cdot c)(H) \equiv 0 \mod |W(H)|$$

*for* $(H) \in \mathscr{H}$. □

8.15 REMARK. According to Theorem 5.8.3 in [tom Dieck 2], *the union* $C$ *of all* $C_{\mathscr{H}}$, which is the subgroup of $\mathbb{Z}^{\Phi(G)}$ free on $\{c_{(H)}, (H) \in \Phi(G)\}$, consists precisely of the *continuous, i.e. locally constant maps* $c : \Phi(G) \to \mathbb{Z}$.

8.16 Also, Corollary 6.3 provides a closed set of congruences satisfied by the image of $F(G)$ in $C \leq \mathbb{Z}^{\Phi(G)}$. Observing that a continuous map $c : \Phi(G) \to \mathbb{Z}$ extends to the set of all subgroups of $G$ as it is a linear combination of the family $\{c_{(H)} = |W(H)|^{-1} \cdot I^*(G/H)\}$, we may assert the following:

**Theorem.** *A continuous map* $c: \Phi(G) \to \mathbb{Z}$ *belongs to the image of the monomorphism* $I^*: F(G) \to C$ *if and only if for every* $(H) \in \Phi(G)$, *the values of* $c$ *satisfy the congruence*

$$\sum_{(K)} |\mathrm{gen}(K/H)|\ |N(H)/(N(H) \cap N(K))| \cdot c(K) \equiv 0 \mod |W(H)|$$

*where the sum is taken over those conjugacy classes in* $G$ *which have a representative* $K$ *containing* $H$ *as a normal subgroup and such that* $K/H$ *is cyclic.*

PROOF. We prove the outstanding assertion by induction on the length of the linear combination $c = \sum_{(L)} n_L(c) \cdot c_{(L)} \in C$.
If its length is zero, $c$ belongs to the image of $I^*$. Otherwise, let $(H) \in \Phi(G)$ be the last index with respect to the ordering "$\preccurlyeq$" among the non-trivial coefficients $n_L = n_L(c) \in \mathbb{Z}$. Due to the congruence for $(H)$, $n_H$ is divisible by $|W(H)|$, for $c_{(L)}$ vanishes on all $(K) \succ (L)$. Thus, $n_H \cdot c_{(H)} = n_H |W(H)|^{-1} \cdot I^*(G/H)$ belongs to the image of $I^*$. Consequently, also the map $c' := c - n_H \cdot c_{(H)} \in C$ satisfies the congruences given. By induction hypothesis, $c'$ is in the image of $I^*$, for there are less non-trivial terms in $c'$ than in $c$. Hence, so is $c = c' + n_H \cdot c_{(H)}$. □

8.17 EXAMPLE. Let us investigate the matrix M when $G$ is a *finite cyclic group* generated by some $g$:

The subgroups of $G$ are uniquely determined by their order. Let $1 = k_0 < k_1 < \cdots < |G|$ be the factors in $|G|$ and let $H_i$ denote the subgroup $\langle g^{k_i} \rangle \leq G$ of index $k_i$. Then $H_j$ is a subgroup of $H_i$ exactly if $k_i$ divides $k_j$. Hence, the *reduced coefficients* $G_i^j = |G/H_i^{H_j}| / |G/H_i|$ of the matrix of $I^*$ are 1 for $k_i \mid k_j$, and 0 otherwise. In the *matrix* $\mathrm{M} = (m_i^j)$ from 8.12, $i$ is the column index, decreasing from left to right, and $j$ the row index, decreasing from top to bottom. For $i < j$, the *recursion formula* 8.13 takes the form

$$-m_i^j = \sum_{\nu} m_i^{\nu}$$

where the sum is taken over $\nu$ with $i \leq \nu < j$ and $k_\nu \mid k_j$. By induction, we see: If some entry $m_i^j$ is non-zero, then $k_i$ divides $k_j$ and in $k_j/k_i$, no prime factor appears twice. Let $P_i(k_j)$ denote the set of primes dividing $k_j/k_i$. Then, outside the diagonal, the *non-trivial entries in* M read

$$m_i^j = (-1)^{|P_i(k_j)|}.$$

Consider the *congruences* 8.14 describing the image of $I^*$: In $(\mathrm{M} \cdot y)(H_j) = \sum_i m_i^j \cdot y(H_i)$ (8.12.), we write $\langle g^{k_i} \rangle$ for $H_i$ and replace $k_i$ by $k_j/(k_j/k_i) = k_j / P_i(k_j)$. The congruences thus take the form

$$\sum_{P(k)} (-1)^{|P(k)|} \cdot y(\langle g^{k/P(k)} \rangle) \equiv 0 \mod k$$

where $k$ runs through the set of all divisors of the order of $g$ and the sum is taken over all subsets $P(k)$ of the set of prime divisors of $k$.

*For $G$ a cyclic $p$-group,* we have $k_i = p^i$, and any $P(k)$ is empty or just the set $\{p\}$. Hence, M has the entries $+1$ on the diagonal, $-1$ just above, and 0 anywhere else. *With $y = I^*(f)$, the congruences read*

$$I(f^{H_i}) - I(f^{H_{i-1}}) \equiv 0 \mod p^i.$$

Because of $f^{H_{i-1}} = f^{\underline{H_i}}$ and $|G/H_i| = p^i$, they are already known from Corollar 5.10. □

**8.18 Corollary.** *Let $E$ be a compact $G$-ENR and consider thereon the multiplication by some element $g \in G$ of finite order. Then, for each $k$ dividing $|g|$, the Hopf indices of $g$ satisfy the congruence*

$$\sum_{P(k)} (-1)^{|P(k)|} \cdot I(g^{k/P(k)}) \equiv 0 \mod k$$

*where the sum is taken over all subsets $P(k)$ of the set of prime factors in $k$ and $k/P(k)$ means $k$ divided by all $p \in P(k)$.*

PROOF. This follows at once from Example 8.17 with $y = I^*(\mathrm{id}_E)$ since for every $g \in G$, $y(g) = \chi(E^g)$ agrees with the Hopf index of the multiplication by $g$ on $E$, as shown in 6.17. □

**8.19** COMMENT. In the borderline case $G = \mathbb{Z}$, there exists a subgroup of index $k$ for every $k \in \mathbb{Z}$. Formally hence, *the iterates of any transformation $g$ on $E$ satisfy the above congruences for all $k \in \mathbb{Z}$* (cf Problem 6.18): This is Theorem 1.1 in [Dold 5].
In fact, this borderline case reduces, generically, to the case of a transformation $g$ of finite order (see [Dold 5], Sections 2 and 5).

## 9. The Fixed Point Index in Equivariant $K$-Theory

**9.1** Let $G$ be a *compact Lie group.* By $K_G(-)$, we denote the *equivariant complex $K$-groups with compact support* (see [Atiyah-Segal] and [Segal 2]): For $X$ a compact $G$-space, $K_G(X)$ is the Grothendieck ring of the isomorphism classes of complex $G$-vector bundles over $X$ and for $X$ locally compact, $K_G(X)$ is the reduced $K_G$-group of the one-point compactification $X^c$ of $X$.
For example, if $p: E \to B$ is a $G$-vector bundle over a compact base, then $K_G(E)$ is the reduced group of the *Thom space* $B^E$ or the relative group of the pair $(P_{\mathbb{R}}(E \oplus \mathbb{R}), P_{\mathbb{R}}(E))$ of projective G-fibre bundles over B. Via the projection $P_{\mathbb{R}}(E \oplus \mathbb{R}) \to B$, $K_G(E)$ becomes a module over the ring $K_G(B)$.

**9.2** As in the non-equivariant case, the elements of the $K_G$-groups of a locally compact $G$-space $X$ are represented by chain complexes of complex $G$-vector bundles over $X$ which have finite length and which are exact outside some compact $G$-subspace of $X$. With $L_G(X)$ denoting suitable equivalence classes of such chain complexes,

$$\chi : L_G \longrightarrow K_G, \quad [0 \to E^1 \to E^2 \to \cdots \to E^n \to 0] \mapsto \sum (-1)^i E^i$$

is a *natural isomorphism of contravariant functors.*

Consider for example a complex $G$-vector bundle $p: E \to B$. By its *Koszul complex* $\Lambda(p)$ one understands the chain complex of exterior algebras $\Lambda^i(p^*(E))$ whose differential is the exterior multiplication by the canonical section of the induced bundle $p^*(E) \to E$. The Koszul complex is exact outside the zero-section in $p$. Therefore, it represents an element $\lambda(p) \in K_G(E)$ provided $B$ is compact. $\lambda(p)$ is called the *Thom class* of the bundle $p$. We also write $\lambda(E)$ if there is no doubt on $p$. The Thom class of the trivial bundle proj: $B \times M \to B$, for example, is the exterior product of the unit element $1_B \in K_G(B)$ with the Thom class $\lambda(M) \in K_G(M)$: $\lambda(\text{proj}) = 1_B \times \lambda(M)$.

Multiplication by $\lambda(p)$ is the *Thom homomorphism for p:*

$$\tau_p := (- \cdot \lambda(p)) : K_G(B) \longrightarrow K_G(E).$$

By means of the functor $L_G$, the Thom homomorphism can also be defined if $B$ is merely locally compact. In any case, it is a *natural isomorphism.*

**9.3** As in the non-equivariant setting, $K_G$ *becomes a multiplicative G-cohomology theory:* For real $G$-modules $M$ and $N$, one sets $\tilde{K}_G^{M-N} := \tilde{K}_G \circ \Sigma^{M \oplus N}$. Thus, $K_G^{-N}(X)$ is $K_G(X \times N)$. The suspension isomorphism $\sigma^M : \tilde{K}_G(X) \to \tilde{K}_G^M(\Sigma^M X) = \tilde{K}_G(\Sigma^{M \oplus M} X)$ is defined as the pointed Thom isomorphism for the trivial bundle $X \times (M \oplus N) \to X$ where $M \oplus M$ carries the complex structure $i \cdot (x, y) := (-y, x)$.

For $X = B^+$, in particular, $\sigma^M$ is the Thom isomorphism $b \mapsto b \times \lambda(M \oplus M)$: $K_G(B) \to K_G(B \times (M \oplus M))$. If we write $s^M \in K_G^M(M) = \tilde{K}_G^M(S^M)$ for $\lambda(M \oplus M) \in K_G(M \oplus M)$, then, as familiar, the suspension isomorphism $\sigma^M$ becomes the exterior multiplication $(- \wedge s^M) : \tilde{K}_G(B^+) \to \tilde{K}_G^M(B^+ \wedge S^M)$.

As $G$ varies in the category $\underline{G}$ of compact Lie groups, we get an *equivariant cohomology theory* $K_{\underline{G}}$. For $i: H \leq G$, the composite

$$K_G(G/H) \xrightarrow{\theta^i} K_H(G/H) \xrightarrow{(i/H)^*} K_H(\mathrm{pt})$$

(see 2.2) simply restricts a $G$-vector bundle over $G/H$ to its fibre over the coset $H$. This is in fact an isomorphism . The inverse homomorphism $\eta_i$ assigns to a complex $H$-module $L$ the $G$-vector bundle $G \times_H L \to G/H$.
The coefficient rings of equivariant $K$-theory are the *complex representation rings* $R(G)$, $G \in \underline{G}$.

9.4 Consider the *Atiyah homomorphism*

$$\alpha : K_G(B) \to K(F \times_G B)\ ,\ [q] \mapsto [\mathrm{id}_F \times_G q]$$

defined for compact $G$-spaces $B$ and $F$. $\alpha$ is the composite of $\mathrm{proj}^*: K_G(B) \to K_G(F \times B)$ with the orbit projection $[q] \mapsto [q/G]$: $K_G(F \times B) \to K(F \times_G B)$. The latter is the homomorphism $\eta_*$ from Remark 4.20 which is an isomorphism when $G$ acts freely on $F$. Hence, if $p$ is a $G$-$\mathrm{ENR}_B$ and $F$ is a free $G$-space such that $\mathrm{id}_F \times_G p$ is a vertical ENR over $F \times_G B$, then, according to Corollary 4.17, $\alpha$ maps the $K_G$-index of a $G$-fixed point situation $f$ in $p$ to the $K$-index of $\mathrm{id}_F \times_G f$.

**Corollary.** *Let $X$ be a compact $G$-space. If $G$ acts freely on $X$, then, for every $H \leq G$,*

$$\chi_{K_G}(G/H) \mapsto \chi_K(X/H \to X/G)$$

*under the Atiyah homomorphism $\alpha$: $R(G) \to K(X/G)$. If there is only one orbit type $(G/H)$ on $X$, then*

$$\chi_{K_{W(H)}}(G/H) \mapsto \chi_K(X \to X/G)$$

*under $\alpha$: $R(W(H)) \to K(X/G)$.*

PROOF. In the first case, $G/H$ is a $G$-ENR and $X \times_G G/H \approx X/H$ is a vertical ENR over $X/G$, being a locally trivial fibre bundle of finite type with fibre an ENR (II, 5.11).
In the second case, $X_H$ is a free $W(H)$-space, $G/H$ a $W(H)$-ENR, and $X_H \times_{W(H)} G/H \approx X$ a vertical ENR over $X_H/W(H) \approx X/G$. □

**9.5** The calculation of the $K_G$-index of a $G$-fixed point situation over a point is based on the following result.

**Proposition.** *If $H \leq G$ is a subgroup of finite index, then the induction*

$$\mathrm{ind}_H^G \colon R(H) \to R(G)$$

*from 4.21 assigns to a complex $H$-module $L$ the induced representation* $\mathrm{Ind}_H^G(L)$ *of $G$.*

PROOF. As shown in 3.11, the stable $G$-transfer of $G/H$ is represented the Pontryagin-Thom construction PT: $S^M \to G/H^N$ for a $G$-embedding of $G/H$ into a $G$-module $M$ with normal bundle $N \to G/H$, followed by the map $G/H^N \to G/H^+ \wedge S^M$ of Thom-spaces induced by the inclusion $N \to G/H \times M$.
So, in order to determine the image of an $H$-module $L$ under our induction $\mathrm{ind}_H^G$, we must pull back to $S^M$, via the PT-construction, the exterior product of the bundle $\eta_i(L) = (G \times_H L \to G/H)$ with $s^M \in \tilde{K}_G^M(S^M)$: We get a multiple of $s^M$ and as claimed, the character function of the factor $\mathrm{ind}_H^G(L) \in R(G)$ proves to be $g \mapsto \Sigma\, \chi_L(\bar{g}^{-1} g \bar{g})$ where the sum is taken over those cosets $\bar{g}H \in G/H$ for which $\bar{g}^{-1} g \bar{g}$ is in $H$:
For, $g \in G$ maps some fibre $L_{\bar{g}H}$ of of $\eta_i(L)$ to itself only if $g$ belongs to the isotropy subgroup $\bar{g}H\bar{g}^{-1}$ of $G$ at $\bar{g}H$. Shifted over to the base fibre $L$, the action of $g$ becomes the multiplication by $\bar{g}^{-1} g \bar{g}$ since the point $[\bar{g}, x] \in L_{\bar{g}H}$ corresponding to some $x \in L$ goes to $[g\bar{g}, x] = [\bar{g}, (\bar{g}^{-1} g \bar{g})x]$. □

**9.6** Hence, according to Corollary 4.22, the $K_G$-characteristic of $G/H$ is the representation of $G$ induced by the trivial $H$-module $\mathbb{C}$, that is the permutation representation $\mathbb{C}(G/H)$. ADD and Corollary 9.4 thus imply:

**Corollary.** *The $K_G$-characteristic of a finite $G$-set $X$ is the permutation representation of $G$ given by $X$:*

$$\chi_{K_G}(X) = \mathbb{C}(X) \in R(G).$$

*Further, let $X$ be a compact $G$-space with a single orbit type $(G/H)$. If $G/H$ is finite, then*

$$\chi_K(X \to X/G) = [X_H \times_{W(H)} \mathbb{C}(G/H) \to X/G] \in K(X/G).$$

*In particular, if G acts freely on X, then, for each subgroup H of finite index in G, we have*

$$\chi_K(X/H \to X/G) = [X \times_G \mathbb{C}(G/H) \to X/G] \in K(X/G).$$

□

**9.7** Applying the sum formula 5.11, we can now calculate the $K_G$-index over a point:

**Theorem.** *The $K_G$-index of a G-fixed point situation f over a point has the character function*

$$I_{K_G}(f): G \to \mathbb{Z}, \; g \mapsto I(f^g).$$

*In particular, the $K_G$-characteristic of a compact G-ENR E is the virtual representation*

$$\chi_{K_G}(E) = \sum (-1)^i H^i(E; \mathbb{C}).$$

Proof. According to the sum formula, $I_{K_G}(f) - I(f^G) \cdot 1_{K_G}$ is a linear combination of the $K_G$-characteristics of the non-trivial orbit types $(G/H)$ with finite $G$-automorphism group $W(H)$. Hence, if $G$ is abelian, there can only contribute orbits $G/H$ of finite length. Evaluated at some point $g \in G$, all terms with $H \not\ni g$ vanish by Corollary 9.6, for then, $g$ acts fixed point free on $G/H$. Consequently, when g happens to be a topological generator of $G$, the character function $I_{K_G}(f)$ takes the value $I(f^G) = I(f^g)$ at $g$.
In the general case, let $i: \langle g \rangle \leq G$ be the closed subgroup generated by $g$. To calculate the trace of $g$, we may regard $I_{K_G}(f)$ as a representation of $\langle g \rangle$. By NAT, $\theta^i(I_{K_G}(f))$ is the $K_{\langle g \rangle}$-index of $i^*(f)$, and hereon, $g$ has the trace $I(f^g)$ as seen before.
In particular, the character of the $K_G$-characteristic of a compact $G$-ENR is the function $g \mapsto \chi(E^g)$. And $\chi(E^g)$ is the Hopf index of the multiplication by $g$ (6.17) which, by the Lefschetz-Hopf trace formula, is the character of the representation $\sum(-1)^i H^i(E; \mathbb{C})$ of $G$ at the point $g$. □

**9.8** Application. For a third time (see 6.2 and 7.8), this shows that, for $G$ finite, the sum of the Hopf indices $I(f^g)$ is divisible by the order of $G$: Indeed, *$\sum_{g \in G} I(f^g)/|G|$ is the dimension of the G-fixed point set of the virtual representation $I_{K_G}(f)$ of G.* For a compact $G$-ENR $E$, Theorem 6.2 thus yields

$$\chi(E/G) = \sum (-1)^i \dim\bigl(H^i(E;\mathbb{C})\bigr)^G.$$

In fact, the existence of a transfer for finite groups even implies $\bigl(H^i(E;\mathbb{C})\bigr)^G \cong H^i(E/G;\mathbb{C})$ ([Bredon], III, 7.2).

**9.9** Another version of a "$K_G$-index" is the *topological index homomorphism*

$$t\text{-ind}_X : K_G(TX) \to R(G)$$

due to [Atiyah-Singer] where $X$ is a compact smooth $G$-manifold with tangent bundle $TX$. If $X$ is a finite $G$-subset of a $G$-module $M$, then $t\text{-ind}_X$ is simply induced by the Pontrjagin-Thom construction corresponding to the embedding of $X$ into the complexification $M_{\mathbb{C}}$ of $M$. In this case hence, it coincides with the $K_G$-transfer of $X$ according to Proposition 3.11. In general, we let $\lambda(TX) \in K_G(TX)$ denote the *real* Thom class of $X$, i.e. the restriction of the Thom class $\lambda(T_{\mathbb{C}}X \to X)$ of the complexified tangent bundle to $TX \subset T_{\mathbb{C}}X$ (see 9.2). Via the $K_G(X)$-module structure on $K_G(TX)$, we get a homomorphism from $K_G(X)$ to $R(G)$, namely $x \mapsto t\text{-ind}_X(x \cdot \lambda(TX))$, which coincides with $t\text{-ind}_X = T^X_{K_G}$ if $X$ is finite.

**Theorem.** *The $K_G$-transfer of a compact $G$-manifold $X$ agrees with the homomorphism*

$$x \mapsto t\text{-ind}_X\bigl(x \cdot \lambda(TX)\bigr) : K_G(X) \to R(G).$$

Proof. To construct $t\text{-ind}_X$, we must embed $X$ as a $G$-subspace into some $G$-module $M$. Let $N_{\mathbb{C}}X$ denote the complexification of the corresponding normal bundle and take the pull-back

$$\begin{array}{ccc} N & \longrightarrow & N_{\mathbb{C}}X \\ \downarrow & & \downarrow \\ TX & \longrightarrow & X \end{array}$$

$N \to TX \to X$ is a real $G$-subbundle of the projection $X \times M_{\mathbb{C}} \to X$. For, if $\tau$ and $\nu$ are the tangent and the normal space of $X$ at some point $x$ - whence $\tau \oplus \nu = M$ - then, along $\tau$, $N$ looks like $\tau \times \nu_{\mathbb{C}} = \tau \oplus (\nu \oplus i\nu)$:

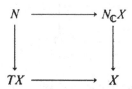

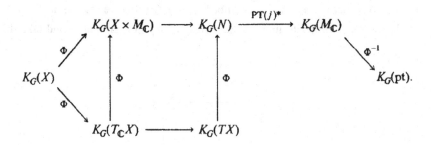

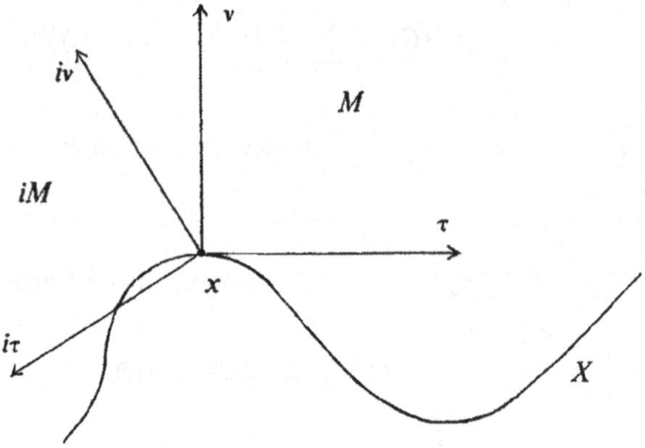


So, since $i(\tau \times v_{\mathbb{C}}) = i(\tau \oplus v) \oplus v$ is isomorphic to the normal space $v \oplus i(\tau \oplus v)$ of $X$ in $M \oplus iM = M_{\mathbb{C}}$, we may think of $N \to X$ as the normal bundle of the $G$-embedding $j: X \to M \to M_{\mathbb{C}}$. The composite of the corresponding Pontrjagin-Thom construction $\mathrm{PT}(j)\colon S^{M_{\mathbb{C}}} \to X^N$ with the Thom isomorphisms for the complex $G$-vector bundles $N \to TX$ and $M_{\mathbb{C}} \to \mathrm{pt}$ now defines the topological index:

$$t\text{-ind}_X\colon K_G(TX) \xrightarrow{\Phi} K_G(N) \xrightarrow{\mathrm{PT}(j)^*} K_G(M_{\mathbb{C}}) \xrightarrow{\Phi^{-1}} K_G(\mathrm{pt}).$$

The multiplication $(-\cdot\lambda(TX))\colon K_G(X) \to K_G(TX)$ is the composite of the Thom isomorphism for the complexified tangent bundle of $X$ with $(TX \to T_{\mathbb{C}}X)^*$ since $X$ is compact (9.2). Hence, $t\text{-ind}_X(-\cdot\lambda(TX))\colon K_G(X) \to K_G(\mathrm{pt})$ is the lower path in the diagram

$$\begin{array}{ccccccc}
 & & K_G(X \times M_{\mathbb{C}}) & \longrightarrow & K_G(N) & \xrightarrow{\mathrm{PT}(j)^*} & K_G(M_{\mathbb{C}}) \\
 & \overset{\Phi}{\nearrow} & \uparrow \Phi & & \uparrow \Phi & & \searrow \Phi^{-1} \\
K_G(X) & & & & & & K_G(\mathrm{pt}). \\
 & \underset{\Phi}{\searrow} & & & & & \\
 & & K_G(T_{\mathbb{C}}X) & \longrightarrow & K_G(TX) & &
\end{array}$$

Naturality reasons ensure that the triangle consisting of Thom isomorphisms commutes, for $X \times M_{\mathbb{C}}$ is the fibrewise product of the bundles $N_{\mathbb{C}}X$ and $T_{\mathbb{C}}X$ over $X$. $N \to TX$ is just the portion $TX \times_X (N_{\mathbb{C}}X)$ of this fibre product over $TX \subset T_{\mathbb{C}}X$. Whence, the square is commutative, too. The map in the upper row, finally, induced by PT($j$) and the inclusion $N \to X \times M_{\mathbb{C}}$, is the $K_G$-transfer of $X$ as shown in 3.11. □

**9.10** REMARK. In particular, *for any compact smooth G-manifold X, t-ind*$_X(\lambda(TX)) \in R(G)$ *is the virtual representation* $\chi_{K_G}(X)$:

$$t\text{-ind}_X(\lambda(TX)) = \sum(-1)^i H^i(X; \mathbb{C}).$$

*This may be derived directly from the Atiyah-Singer index formula:*

We consider finite chain complexes of *G-differential operators over X*, $d_i : \Gamma(p_i) \to \Gamma(p_{i+1})$, $d_{i+1} d_i = 0$. In detail: When $p$ is a smooth complex $G$-vector bundle over $X$, let $\Gamma(p)$ denote the $G$-module of its smooth cross section. Given two such section modules and a $G$-homomorphism $d$ inbetween which, in local coordinates, takes the form of a usual differential operator, then $d$ is what is understood by a $G$-differential operator on $X$.
Any such chain complex $(d) = (d_i)$ represents an element in $R(G)$, namely $\chi(d) = \Sigma(-1)^i \Gamma(p_i)$ (see 9.2). The local characteristic polynomials of the differential operators in $(d)$ assemble to form a chain complex of complex $G$-vector bundles over the dual of $TX$, the so-called *symbol of* $(d)$:

$$\sigma(d_i) : (T^*X \to X)^*(p_i) \longrightarrow (T^*X \to X)^*(p_{i+1}).$$

$(d)$ is called *elliptic* if $\sigma(d)$ is exact outside the zero-section. If $X$ is now compact - which was not required so far - then $\sigma(d)$ represents an element of $K_G(T^*X)$ and the *index formula of Atiyah and Singer* says

$$t\text{-ind}_X \circ \chi \circ \delta^* (\sigma(d)) = \chi(d) \quad \text{in short} \quad t\text{-ind}_X[\sigma(d)] = [d].$$

$\delta: TX \to T^*X$ is the $G$-isomorphism induced by some $G$-invariant metric on $X$.

Consider for example the $\mathbb{C}$-valued *de Rham complex* $\Omega(X)$ of a compact smooth $G$-manifold. This is the chain complex built by the smooth sections of the alternating forms on $T_{\mathbb{C}}X$, i.e. of the exterior algebras of the dual $T_{\mathbb{C}}{}^*X$. Its symbol is the Koszul complex of the co-tangent bundle $T_{\mathbb{C}}{}^*X \to X$ pulled back to $T^*X$ (see 9.2). Hence, $\Omega(X)$ is elliptic and

$\delta^*(\sigma(\Omega X))$ represents the real Thom class $\lambda(TX)$, for the two bundles $T_{\mathbb{C}}X$ and $T_{\mathbb{C}}{}^*X$ are isomorphic by virtue of $\delta_{\mathbb{C}}$.
The index formula thus yields $t\text{-ind}_X(\lambda(TX)) = \chi(\Omega X)$. Substituting $\Omega X$ by its cohomology complex, the assertion follows from the de Rham theorem $H^i(\Omega X) \cong H^i(X; \mathbb{C})$. □

**9.11** COMMENT. According to Proposition 9.5, our induction homomorphism

$$\text{ind}_H^G : R(H) \xrightarrow{\eta_i} K_G(G/H) \xrightarrow{T_{K_G}^{G/H}} R(G)$$

extends the construction of the induced representation familiar for finite groups to inclusions $i: H \leq G$ of compact Lie groups. In [Segal 1], one finds a homomorphism

$$i_*: R(H) \to R(G)$$

which does the same. As detailed in [Nishida], $i_*(L)$ equals $t\text{-ind}_{G/H}(\eta_i(L) \cdot \lambda(T(G/H)))$ for any $H$-module $L$. Hence, $i_*$ *coincides with* $\text{ind}_H^G$ in view of Theorem 9.9. Without much ado, we can deduce $i_* = \text{ind}_H^G$ on the image of $\theta^i: R(G) \to R(H)$:

On the one hand, we know from 4.21 that $\text{ind}_H^G \circ \theta^i$ is multiplication by $\chi_{K_G}(G/H)$. On the other hand, $i_*$ is $R(G)$-multiplicative, hence $i_* \circ \theta^i$ is multiplication by $i_*(1)$ and by definition, $i_*(1)$ is represented by the section complex of the exterior algebras of $T_{\mathbb{C}}{}^*(G/H)$ over $G/H$. It is even sufficient to take smooth sections since these are dense in the space of all (see [Segal 1]): But the smooth sections just form the $\mathbb{C}$-valued de Rham complex $\Omega(G/H)$ which was shown above to represent $\chi_{K_G}(G/H)$. □

# APPENDIX

For completeness, we enclose from [Ulrich] some results on proper maps which have been used repeatedly in the Sections II, 1 and III, 1 when we were discussing fundamental properties of $ENR_B$s and index morphisms. They are well-known for locally compact spaces and we want to generalize them to compactly generated spaces.

## A. On Proper Maps and Compactly Generated Spaces

**A.1** Since all spaces considered so far were Hausdorff, *we demand that a compactly generated space be Hausdorff.* This will keep our results simple.

**A.2** We follow [Bourbaki] to define proper maps:

**Definition.** A map $f: X \to Y$ of topological spaces is called *proper*, if $f \times \mathrm{id}: X \times Z \to Y \times Z$ is a closed map for any space Z. We say $f$ is *locally proper* if any point in $X$ has a neighbourhood $U$ such that $f \mid U: U \to Y$ is proper.

**A.3** The following criterion is well-known:

**Proposition.** *A map $f: X \to Y$ is proper if and only if $f$ is closed and $f^{-1}(y)$ is quasicompact for all $y \in Y$.* □

**A.4** We now come to the results required.

**Proposition.** *If a map $f: X \to Y$ is proper, then $f^{-1}(K)$ is quasicompact for all quasicompact subspaces $K \subset Y$.*
*The converse holds if $Y$ is compactly generated. In this case, the topology of $X$, too, is compactly generated provided $X$ is a Hausdorff space.*

PROOF. Suppose first that $f$ is a proper map. Take $K \subset Y$ quasicompact and let $\{V_\lambda\}$ be an open covering of $f^{-1}(K)$. For each $k \in K$, select a finite subfamily of $\{V_\lambda\}$ covering $f^{-1}(k)$ and denote its union by $V^k$. Then $U^k := \{y \in Y: f^{-1}(y) \subset V^k\}$ is an open neighbourhood

of $k$ in $Y$ because $f$ is a closed map. Now, a finite number of the $U^k$s suffices to cover $K$ and $f^{-1}(U^k) \subset V^k$ implies that the corresponding finite subfamily of $\{V^k\}$ covers $f^{-1}(K)$.
For the converse, we have to show that $f$ is a closed map. So let $A \subset X$ be closed and consider a compactum $K \subset Y$. Then $f(A) \cap K = f\big(A \cap f^{-1}(K)\big)$ is compact because $f^{-1}(K)$ is quasicompact.
To show finally that $X$ is compactly generated, let $A$ be a subspace of $X$ such that $A \cap C$ is closed and hence compact for all compacta $C \subset X$. Thus, by assumption, $A \cap f^{-1}(K)$ is compact if $K \subset Y$ is so. Whence, $f$ is proper on $A$. The outstanding assertion is proved separately. □

**A.5 Lemma.** *Let $f: X \to Y$ be a map from a Hausdorff space to a compactly generated space. If $f$ is proper on a subspace $A \subset X$, then $A$ is closed.*

PROOF. Consider some $x \notin A$ and set $z := f(x)$. Let $f': A \to Y$ denote the restriction of $f$. By assumption, $f'^{-1}(z)$ is compact. Hence, there exist disjoint open neighbourhoods $U$ of $x$ and $V$ of $f'^{-1}(z)$ in $X$. Then $W := \{y \in Y : f'^{-1}(y) \subset V\}$ is a neighbourhood of $z$ because $f'$ is a closed map. Hence, $f^{-1}(W) \cap U$ is a neighbourhood of $x$ in $X$ completely outside $A$. □

**A.6** With that, the following special characterization given in [Dold 3], 1.3, generalizes as well:

**Proposition.** *Let $B$ be a paracompact space whose topology is compactly generated. Then the projection $B \times \mathbb{R}^n \to B$ is proper on some subspace $A \subset B \times \mathbb{R}^n$ if and only if $A$ is closed and fibrewise uniformly bounded.* □

**A.7** Besides, we have used in Section III, 1 that $B \times \mathbb{R}^n$ is compactly generated if $B$ is. More generally, we can show:

**Proposition.** *Let $X$ be a compactly generated space. If $Y$ is locally compact, then $X \times Y$ is compactly generated.*

PROOF. If $Y$ is even compact, we know from A.4 that the projection $X \times Y \to X$ is a proper map, which implies that $X \times Y$ is compactly generated.
Otherwise, let $O \subset X \times Y$ be so that $O \cap C$ is open in $C$ for all compacta $C \subset X \times Y$. Then, if $K \subset Y$ is compact, $O \cap (X \times K)$ is open in $X \times K$, for we know that $X \times K$ is compactly

generated. But since $Y$ is locally compact, this means that $O$ is a neighbourhood of all of its points. □

# Bibliography


Adams, J.F.:
[1] *Infinite Loop Spaces.* Annals of Math. Studies **90**. Princeton University Press 1978.
[2] Graeme Segal's Burnside ring conjecture. *Bull. Amer. Math. Soc.* 6 (1982), 201-210.

Addis, D.:
A strong regularity condition on mappings. *General Top. and its Applic.* 2 (1972), 199-213.

Allaud, G. - Fadell, E.:
A fibre homotopy extension theorem. *Trans. Amer. Math. Soc.* **104** (1962), 239-251.

Atiyah, M.F. - Segal, G.B.:
*Equivariant K-Theory.* Lecture Notes. University of Warwick 1965.

Atiyah, M.F. - Singer, I.M.:
The Index of elliptic operators, I. *Annals of Math.* **87** (1968), 484-530.

Bierstone, E.:
The equivariant covering homotopy property for differentiable $G$-fibre bundles.
*J. of Diff. Geom.* **8** (1973), 615-622.

Birkhoff, G.:
*Lattice Theory.* Amer. Math. Soc. Colloquium Publications **XXV**.
Providence, Rhode Island 1948.

Bredon, G.E.:
*Introduction to Compact Transformation Groups.* Academic Press, New York-London 1972.

Bröcker, T. - Jänich, K.:
*Einführung in die Differentialtopologie.* Heidelberger Taschenbücher **143**.
Springer, Berlin-Heidelberg-New York 1973.

Brown, K.S.:
Complete Euler characteristics and fixed point theory.
*J. of Pure and Appl. Algebra* **24** (1982), 103-121.

Clapp de Prieto, M.:
*Dualität in der Kategorie der Spektren von Ex-Räumen.* Dissertation. Heidelberg 1979.

Conner, P.E. - Floyd, E.E.:
*Differentiable Periodic Maps.* Ergebnisse der Mathematik und ihrer Grenzgebiete **33**.
Springer, Berlin-Heidelberg-New York 1964.

tom Dieck, T.:
[1] Faserbündel mit Gruppenoperation. *Archiv d. Math.* **20** (1969), 136-143.
[2] *Transformation Groups and Representation Theory.* Lecture Notes in Mathematics 766.
Springer, Berlin-Heidelberg-New York 1979.

tom Dieck, T. - Kamps, K.H. - Puppe, D. [DKP]:
*Homotopietheorie.* Lecture Notes in Math. **157**. Springer, Berlin-Heidelberg-New York 1970.

Dold, A.:
[1] Die Homotopieerweiterungseigenschaft ist eine lokale Eigenschaft.
*Invent. Math.* 6 (1968), 185-189.
[2] The fixed point index of fibre preserving maps. *Invent. Math.* **25** (1974), 281-297.
[3] The fixed point transfer of fibre preserving maps. *Math. Z.* **148** (1976), 215-244.
[4] *Elementare Theorie des Abbildungsgrades und des Fixpunktindexes.*
Vorlesungsausarbeitung. Heidelberg 1978.
[5] Fixed point indices of iterated maps. *Invent. Math.* 74 (1983), 419-435.

Dold, A. - Puppe, D.:
Duality, trace and transfer.
*Proc. of the Internat. Conf. on Geom. Topology* **Warszawa** (1980), 81-102.

Dowker, C.H.:
An embedding theorem for paracompact metric spaces. *Duke Math. J.* 14 (1947), 639-645.

Ehresmann, C.:
Sur les espaces fibrés différentiables. *CR Acad. Sci. Paris* **224** (1947), 1611.

Engelking, R.:
*General Topology.* Monografie Matematyczne **60**. Warszawa 1977.

Ferry, S.:
Strongly regular mappings with compact ANR fibres are Hurewicz fiberings. *Pacific J. of Math.* **75** (1978), 373-382.
Gottlieb, D.H.:
The Lefschetz number and Borsuk-Ulam theorems. *Preprint.* Purdue University 1982.
Hauschild, H.:
[1] Äquivariante Transversalität und äquivariante Bordismentheorie. *Archiv d. Math.* **26** (1975), 536-546.
[2] Zerspaltung äquivarianter Homotopiemengen. *Math. Ann.* **230** (1977), 279-292.
Hopf, H.:
Über den Rang geschlossener Liescher Gruppen. *Comment. Math. Helv.* **13** (1940), 119-143.
Hu, S.T.:
*Theory of Retracts.* Wayne State University Press. Detroit 1965.
Hurewicz, W.:
On the concept of a fibre space. *Proc. Nat. Acad. Sci.* **41** (1955), 956-961.
Husemoller, D.:
*Fibre Bundles.* Second Edition. Graduate Texts in Math. **20**. Springer, Berlin-Heidelberg-New York 1975.
Jänich, K.:
*Differenzierbare G-Mannigfaltigkeiten.* Lecture Notes in Mathematics **59**. Springer, Berlin-Heidelberg-New York 1968.
Jaworowski, J.:
Extension of $G$-maps and euclidean $G$-retracts. *Math. Z.* **146** (1976), 143-148.
Kosniowski, C.:
Equivariant cohomology and stable cohomotopy. *Math. Ann.* **210** (1974), 83-104.
Koszul, J.L.:
Sur certains groupes de transformation de Lie. *Coll. Int. CNRS de Géom. Diff.* **52** (1953), 137-142.
Laitinen, E.:
*On the Burnside Ring and Stable Cohomotopy of a Finite Group.* Aarhus University Preprint Series 14 1977/78.
Lashof, R.K.:
[1] The equivariant extension theorem. *Proc. Amer. Math. Soc.* **83** (1983), 138-140.
[2] Equivariant bundles. *Illinois J. of Math.* **26** (1982), 257-271.
Mahammed, N. - Piccinini, R. - Stuter, U.:
*Some Applications of Topological K-Theory.* Math. Studies **45**. North-Holland, Amsterdam 1980.
Michael, E.:
A note on paracompact spaces. *Proc. Amer. Math. Soc.* **4** (1953), 831-838.
Morita, K.:
[1] On the dimension of product spaces. *Amer. J. of Math.* **75** (1953), 205-223.
[2] Products of normal spaces with metric spaces. *Math. Ann.* **154** (1964), 365-382.
Mostow, G.D.:
Equivariant embedding in euclidean space. *Annals of Math.* **65** (1957), 432-446.
Nagami, K.:
*Dimension Theory.* Pure and Applied Mathematics. Academic Press, New York-London 1970.
Nagata, J.:
*Modern General Topology.* Bibliotheca Mathematica VII. North-Holland, Amsterdam 1968.
Nishida, G.:
The transfer homomorphism in equivariant generalized cohomology theories. *J. of Math. Kyoto Univ.* **18** (1978), 435-451.
Palais, R.S.:
[1] Embedding of compact differentiable transformation groups in orthogonal representations. *J. of Math. and Mech.* **6** (1957), 673-678.
[2] *The Classification of G-Spaces.* Memoirs Amer. Math. Soc. **36**. Providence, Rhode Island 1960.
Prieto, C.:
Coincidence index for fibre preserving maps. *Manusc. Math.* **47** (1984), 233-249.
Samelson, H. - Hopf, H.:
Ein Satz über die Wirkungsräume geschlossener Liescher Gruppen. *Comment. Math. Helv.* **13** (1940), 240-251.

Segal, G.B.:
[1] The representation ring of a compact Lie group. *Publ. Math. IHES* **34** (1968), 113-128.
[2] Equivariant *K*-theory. *Publ. Math. IHES* **34** (1968), 129-151.
[3] Equvariant stable homotopy theory. *Actes Congres Intern. Math.* **2** (1970), 59-63.
[4] Configuration spaces and iterated loop spaces. *Invent. Math.* 21 (1973), 213-221.

Serre, J.-P.:
*Représentations Linéaires des Groupes Finis.* 2ème édition. Hermann, Paris 1967.

Shanahan, P.:
*The Atiyah-Singer Index Theorem.* Lecture Notes in Math. **638.**
Springer, Berlin-Heidelberg-New York 1978.

Spanier, E.H.:
*Algebraic Topology.* McGraw-Hill, New York 1966.

Steenrod, N.:
*The Topology of Fibre Bundles.* Princeton University Press 1951.

Ulrich, H.:
*Der Fixpunktindex fasernweiser Abbildungen.* Diplomarbeit. Heidelberg 1976.

Ungar, G.S.:
The slicing structure property. *Pac. J. Math.* **30** (1969), 549-553.

Weil, A.:
Démonstration topologique d'un théorème fondamental de Cartan.
*CR Acad. Sci. Paris* **200** (1935), 518-520.

# Index of Notation and Terminology

Items are kept in logical groups referred to by chapter and section number.

## LECTURE NOTES IN MATHEMATICS

Edited by A. Dold and B. Eckmann

### Some general remarks on the publication of monographs and seminars

In what follows all references to monographs, are applicable also to multiauthorship volumes such as seminar notes.

§1. Lecture Notes aim to report new developments - quickly, informally, and at a high level. Monograph manuscripts should be reasonably self-contained and rounded off. Thus they may, and often will, present not only results of the author but also related work by other people. Furthermore, the manuscripts should provide sufficient motivation, examples and applications. This clearly distinguishes Lecture Notes manuscripts from journal articles which normally are very concise. Articles intended for a journal but too long to be accepted by most journals, usually do not have this "lecture notes" character. For similar reasons it is unusual for Ph.D. theses to be accepted for the Lecture Notes series.

Experience has shown that English language manuscripts achieve a much wider distribution.

§2. Manuscripts or plans for Lecture Notes volumes should be submitted either to one of the series editors or to Springer-Verlag, Heidelberg. These proposals are then refereed. A final decision concerning publication can only be made on the basis of the complete manuscripts, but a preliminary decision can usually be based on partial information: a fairly detailed outline describing the planned contents of each chapter, and an indication of the estimated length, a bibliography, and one or two sample chapters - or a first draft of the manuscript. The editors will try to make the preliminary decision as definite as they can on the basis of the available information.

§3. Lecture Notes are printed by photo-offset from typed copy delivered in camera-ready form by the authors. Springer-Verlag provides technical instructions for the preparation of manuscripts, and will also, on request, supply special staionery on which the prescribed typing area is outlined. Careful preparation of the manuscripts will help keep production time short and ensure satisfactory appearance of the finished book. Running titles are not required; if however they are considered necessary, they should be uniform in appearance. We generally advise authors not to start having their final manuscripts specially tpyed beforehand. For professionally typed manuscripts, prepared on the special stationery according to our instructions, Springer-Verlag will, if necessary, contribute towards the typing costs at a fixed rate.

The actual production of a Lecture Notes volume takes 6-8 weeks.

.../...

§4. Final manuscripts should contain at least 100 pages of mathematical text and should include
- a table of contents
- an informative introduction, perhaps with some historical remarks. It should be accessible to a reader not particularly familiar with the topic treated.
- a subject index; this is almost always genuinely helpful for the reader.

§5. Authors receive a total of 50 free copies of their volume, but no royalties. They are entitled to purchase further copies of their book for their personal use at a discount of 33.3 %, other Springer mathematics books at a discount of 20 % directly from Springer-Verlag.

Commitment to publish is made by letter of intent rather than by signing a formal contract. Springer-Verlag secures the copyright for each volume.

Vol. 1173: H. Delfs, M. Knebusch, Locally Semialgebraic Spaces. XVI, 329 pages. 1985.

Vol. 1174: Categories in Continuum Physics, Buffalo 1982. Seminar. Edited by F.W. Lawvere and S.H. Schanuel. V, 126 pages. 1986.

Vol. 1175: K. Mathiak, Valuations of Skew Fields and Projective Hjelmslev Spaces. VII, 116 pages. 1986.

Vol. 1176: R.R. Bruner, J.P. May, J.E. McClure, M. Steinberger, $H_\infty$ Ring Spectra and their Applications. VII, 388 pages. 1986.

Vol. 1177: Representation Theory I. Finite Dimensional Algebras. Proceedings, 1984. Edited by V. Dlab, P. Gabriel and G. Michler. XV, 340 pages. 1986.

Vol. 1178: Representation Theory II. Groups and Orders. Proceedings, 1984. Edited by V. Dlab, P. Gabriel and G. Michler. XV, 370 pages. 1986.

Vol. 1179: Shi J.-Y. The Kazhdan-Lusztig Cells in Certain Affine Weyl Groups. X, 307 pages. 1986.

Vol. 1180: R. Carmona, H. Kesten, J.B. Walsh, École d'Été de Probabilités de Saint-Flour XIV – 1984. Édité par P.L. Hennequin. X, 438 pages. 1986.

Vol. 1181: Buildings and the Geometry of Diagrams, Como 1984. Seminar. Edited by L. Rosati. VII, 277 pages. 1986.

Vol. 1182: S. Shelah, Around Classification Theory of Models. VII, 279 pages. 1986.

Vol. 1183: Algebra, Algebraic Topology and their Interactions. Proceedings, 1983. Edited by J.-E. Roos. XI, 396 pages. 1986.

Vol. 1184: W. Arendt, A. Grabosch, G. Greiner, U. Groh, H.P. Lotz, U. Moustakas, R. Nagel, F. Neubrander, U. Schlotterbeck, One-parameter Semigroups of Positive Operators. Edited by R. Nagel. X, 460 pages. 1986.

Vol. 1185: Group Theory, Beijing 1984. Proceedings. Edited by Tuan H.F. V, 403 pages. 1986.

Vol. 1186: Lyapunov Exponents. Proceedings, 1984. Edited by L. Arnold and V. Wihstutz. VI, 374 pages. 1986.

Vol. 1187: Y. Diers, Categories of Boolean Sheaves of Simple Algebras. VI, 168 pages. 1986.

Vol. 1188: Fonctions de Plusieurs Variables Complexes V. Séminaire, 1979–85. Edité par François Norguet. VI, 306 pages. 1986.

Vol. 1189: J. Lukeš, J. Malý, L. Zajíček, Fine Topology Methods in Real Analysis and Potential Theory. X, 472 pages. 1986.

Vol. 1190: Optimization and Related Fields. Proceedings, 1984. Edited by R. Conti, E. De Giorgi and F. Giannessi. VIII, 419 pages. 1986.

Vol. 1191: A.R. Its, V.Yu. Novokshenov, The Isomonodromic Deformation Method in the Theory of Painlevé Equations. IV, 313 pages. 1986.

Vol. 1192: Equadiff 6. Proceedings, 1985. Edited by J. Vosmansky and M. Zlámal. XXIII, 404 pages. 1986.

Vol. 1193: Geometrical and Statistical Aspects of Probability in Banach Spaces. Proceedings, 1985. Edited by X. Fernique, B. Heinkel, M.B. Marcus and P.A. Meyer. IV, 128 pages. 1986.

Vol. 1194: Complex Analysis and Algebraic Geometry. Proceedings, 1985. Edited by H. Grauert. VI, 235 pages. 1986.

Vol. 1195: J.M. Barbosa, A.G. Colares, Minimal Surfaces in $\mathbb{R}^3$. X, 124 pages. 1986.

Vol. 1196: E. Casas-Alvero, S. Xambó-Descamps, The Enumerative Theory of Conics after Halphen. IX, 130 pages. 1986.

Vol. 1197: Ring Theory. Proceedings, 1985. Edited by F.M.J. van Oystaeyen. V, 231 pages. 1986.

Vol. 1198: Séminaire d'Analyse, P. Lelong – P. Dolbeault – H. Skoda. Seminar 1983/84. X, 260 pages. 1986.

Vol. 1199: Analytic Theory of Continued Fractions II. Proceedings, 1985. Edited by W.J. Thron. VI, 299 pages. 1986.

Vol. 1200: V.D. Milman, G. Schechtman, Asymptotic Theory of Finite Dimensional Normed Spaces. With an Appendix by M. Gromov. VIII, 156 pages. 1986.

Vol. 1201: Curvature and Topology of Riemannian Manifolds. Proceedings, 1985. Edited by K. Shiohama, T. Sakai and T. Sunada. VII, 336 pages. 1986.

Vol. 1202: A. Dür, Möbius Functions, Incidence Algebras and Power Series Representations. XI, 134 pages. 1986.

Vol. 1203: Stochastic Processes and Their Applications. Proceedings, 1985. Edited by K. Itô and T. Hida. VI, 222 pages. 1986.

Vol. 1204: Séminaire de Probabilités XX, 1984/85. Proceedings. Edité par J. Azéma et M. Yor. V, 639 pages. 1986.

Vol. 1205: B.Z. Moroz, Analytic Arithmetic in Algebraic Number Fields. VII, 177 pages. 1986.

Vol. 1206: Probability and Analysis, Varenna (Como) 1985. Seminar. Edited by G. Letta and M. Pratelli. VIII, 280 pages. 1986.

Vol. 1207: P.H. Bérard, Spectral Geometry: Direct and Inverse Problems. With an Appendix by G. Besson. XIII, 272 pages. 1986.

Vol. 1208: S. Kaijser, J.W. Pelletier, Interpolation Functors and Duality. IV, 167 pages. 1986.

Vol. 1209: Differential Geometry, Peñíscola 1985. Proceedings. Edited by A.M. Naveira, A. Ferrández and F. Mascaró. VIII, 306 pages. 1986.

Vol. 1210: Probability Measures on Groups VIII. Proceedings, 1985. Edited by H. Heyer. X, 386 pages. 1986.

Vol. 1211: M.B. Sevryuk, Reversible Systems. V, 319 pages. 1986.

Vol. 1212: Stochastic Spatial Processes. Proceedings, 1984. Edited by P. Tautu. VIII, 311 pages. 1986.

Vol. 1213: L.G. Lewis, Jr., J.P. May, M. Steinberger, Equivariant Stable Homotopy Theory. IX, 538 pages. 1986.

Vol. 1214: Global Analysis – Studies and Applications II. Edited by Yu.G. Borisovich and Yu.E. Gliklikh. V, 275 pages. 1986.

Vol. 1215: Lectures in Probability and Statistics. Edited by G. del Pino and R. Rebolledo. V, 491 pages. 1986.

Vol. 1216: J. Kogan, Bifurcation of Extremals in Optimal Control. VIII, 106 pages. 1986.

Vol. 1217: Transformation Groups. Proceedings, 1985. Edited by S. Jackowski and K. Pawalowski. X, 396 pages. 1986.

Vol. 1218: Schrödinger Operators, Aarhus 1985. Seminar. Edited by E. Balslev. V, 222 pages. 1986.

Vol. 1219: R. Weissauer, Stabile Modulformen und Eisensteinreihen. III, 147 Seiten. 1986.

Vol. 1220: Séminaire d'Algèbre Paul Dubreil et Marie-Paule Malliavin. Proceedings, 1985. Edité par M.-P. Malliavin. IV, 200 pages. 1986.

Vol. 1221: Probability and Banach Spaces. Proceedings, 1985. Edited by J. Bastero and M. San Miguel. XI, 222 pages. 1986.

Vol. 1222: A. Katok, J.-M. Strelcyn, with the collaboration of F. Ledrappier and F. Przytycki, Invariant Manifolds, Entropy and Billiards; Smooth Maps with Singularities. VIII, 283 pages. 1986.

Vol. 1223: Differential Equations in Banach Spaces. Proceedings, 1985. Edited by A. Favini and E. Obrecht. VIII, 299 pages. 1986.

Vol. 1224: Nonlinear Diffusion Problems, Montecatini Terme 1985. Seminar. Edited by A. Fasano and M. Primicerio. VIII, 188 pages. 1986.

Vol. 1225: Inverse Problems, Montecatini Terme 1986. Seminar. Edited by G. Talenti. VIII, 204 pages. 1986.

Vol. 1226: A. Buium, Differential Function Fields and Moduli of Algebraic Varieties. IX, 146 pages. 1986.

Vol. 1227: H. Helson, The Spectral Theorem. VI, 104 pages. 1986.

Vol. 1228: Multigrid Methods II. Proceedings, 1985. Edited by W. Hackbusch and U. Trottenberg. VI, 336 pages. 1986.

Vol. 1229: O. Bratteli, Derivations, Dissipations and Group Actions on $C^*$-algebras. IV, 277 pages. 1986.

Vol. 1230: Numerical Analysis. Proceedings, 1984. Edited by J.-P. Hennart. X, 234 pages. 1986.

Vol. 1231: E.-U. Gekeler, Drinfeld Modular Curves. XIV, 107 pages. 1986.

Vol. 1232: P.C. Schuur, Asymptotic Analysis of Soliton Problems. VIII, 180 pages. 1986.

Vol. 1233: Stability Problems for Stochastic Models. Proceedings, 1985. Edited by V.V. Kalashnikov, B. Penkov and V.M. Zolotarev. VI, 223 pages. 1986.

Vol. 1234: Combinatoire énumérative. Proceedings, 1985. Edité par G. Labelle et P. Leroux. XIV, 387 pages. 1986.

Vol. 1235: Séminaire de Théorie du Potentiel, Paris, No. 8. Directeurs: M. Brelot, G. Choquet et J. Deny. Rédacteurs: F. Hirsch et G. Mokobodzki. III, 209 pages. 1987.

Vol. 1236: Stochastic Partial Differential Equations and Applications. Proceedings, 1985. Edited by G. Da Prato and L. Tubaro. V, 257 pages. 1987.

Vol. 1237: Rational Approximation and its Applications in Mathematics and Physics. Proceedings, 1985. Edited by J. Gilewicz, M. Pindor and W. Siemaszko. XII, 350 pages. 1987.

Vol. 1238: M. Holz, K.-P. Podewski and K. Steffens, Injective Choice Functions. VI, 183 pages. 1987.

Vol. 1239: P. Vojta, Diophantine Approximations and Value Distribution Theory. X, 132 pages. 1987.

Vol. 1240: Number Theory, New York 1984–85. Seminar. Edited by D.V. Chudnovsky, G.V. Chudnovsky, H. Cohn and M.B. Nathanson. V, 324 pages. 1987.

Vol. 1241: L. Gårding, Singularities in Linear Wave Propagation. III, 125 pages. 1987.

Vol. 1242: Functional Analysis II, with Contributions by J. Hoffmann-Jørgensen et al. Edited by S. Kurepa, H. Kraljević and D. Butković. VII, 432 pages. 1987.

Vol. 1243: Non Commutative Harmonic Analysis and Lie Groups. Proceedings, 1985. Edited by J. Carmona, P. Delorme and M. Vergne. V, 309 pages. 1987.

Vol. 1244: W. Müller, Manifolds with Cusps of Rank One. XI, 158 pages. 1987.

Vol. 1245: S. Rallis, L-Functions and the Oscillator Representation. XVI, 239 pages. 1987.

Vol. 1246: Hodge Theory. Proceedings, 1985. Edited by E. Cattani, F. Guillén, A. Kaplan and F. Puerta. VII, 175 pages. 1987.

Vol. 1247: Séminaire de Probabilités XXI. Proceedings. Edité par J. Azéma, P.A. Meyer et M. Yor. IV, 579 pages. 1987.

Vol. 1248: Nonlinear Semigroups, Partial Differential Equations and Attractors. Proceedings, 1985. Edited by T.L. Gill and W.W. Zachary. IX, 185 pages. 1987.

Vol. 1249: I. van den Berg, Nonstandard Asymptotic Analysis. IX, 187 pages. 1987.

Vol. 1250: Stochastic Processes – Mathematics and Physics II. Proceedings 1985. Edited by S. Albeverio, Ph. Blanchard and L. Streit. VI, 359 pages. 1987.

Vol. 1251: Differential Geometric Methods in Mathematical Physics. Proceedings, 1985. Edited by P.L. García and A. Pérez-Rendón. VII, 300 pages. 1987.

Vol. 1252: T. Kaise, Représentations de Weil et $GL_2$ Algèbres de division et $GL_n$. VII, 203 pages. 1987.

Vol. 1253: J. Fischer, An Approach to the Selberg Trace Formula via the Selberg Zeta-Function. III, 184 pages. 1987.

Vol. 1254: S. Gelbart, I. Piatetski-Shapiro, S. Rallis. Explicit Constructions of Automorphic L-Functions. VI, 152 pages. 1987.

Vol. 1255: Differential Geometry and Differential Equations. Proceedings, 1985. Edited by C. Gu, M. Berger and R.L. Bryant. XII, 243 pages. 1987.

Vol. 1256: Pseudo-Differential Operators. Proceedings, 1986. Edited by H.O. Cordes, B. Gramsch and H. Widom. X, 479 pages. 1987.

Vol. 1257: X. Wang, On the C*-Algebras of Foliations in the Plane. V, 165 pages. 1987.

Vol. 1258: J. Weidmann, Spectral Theory of Ordinary Differential Operators. VI, 303 pages. 1987.

Vol. 1259: F. Cano Torres, Desingularization Strategies for Three-Dimensional Vector Fields. IX, 189 pages. 1987.

Vol. 1260: N.H. Pavel, Nonlinear Evolution Operators and Semigroups. VI, 285 pages. 1987.

Vol. 1261: H. Abels, Finite Presentability of S-Arithmetic Groups. Compact Presentability of Solvable Groups. VI, 178 pages. 1987.

Vol. 1262: E. Hlawka (Hrsg.), Zahlentheoretische Analysis II. Seminar, 1984–86. V, 158 Seiten. 1987.

Vol. 1263: V.L. Hansen (Ed.), Differential Geometry. Proceedings, 1985. XI, 288 pages. 1987.

Vol. 1264: Wu Wen-tsün, Rational Homotopy Type. VIII, 219 pages. 1987.

Vol. 1265: W. Van Assche, Asymptotics for Orthogonal Polynomials. VI, 201 pages. 1987.

Vol. 1266: F. Ghione, C. Peskine, E. Sernesi (Eds.), Space Curves. Proceedings, 1985. VI, 272 pages. 1987.

Vol. 1267: J. Lindenstrauss, V.D. Milman (Eds.), Geometrical Aspects of Functional Analysis. Seminar. VII, 212 pages. 1987.

Vol. 1268: S.G. Krantz (Ed.), Complex Analysis. Seminar, 1986. VII, 195 pages. 1987.

Vol. 1269: M. Shiota, Nash Manifolds. VI, 223 pages. 1987.

Vol. 1270: C. Carasso, P.-A. Raviart, D. Serre (Eds.), Nonlinear Hyperbolic Problems. Proceedings, 1986. XV, 341 pages. 1987.

Vol. 1271: A.M. Cohen, W.H. Hesselink, W.L.J. van der Kallen, J.R. Strooker (Eds.), Algebraic Groups Utrecht 1986. Proceedings. XII, 284 pages. 1987.

Vol. 1272: M.S. Livšic, L.L. Waksman, Commuting Nonselfadjoint Operators in Hilbert Space. III, 115 pages. 1987.

Vol. 1273: G.-M. Greuel, G. Trautmann (Eds.), Singularities, Representation of Algebras, and Vector Bundles. Proceedings, 1985. XIV, 383 pages. 1987.

Vol. 1274: N.C. Phillips, Equivariant K-Theory and Freeness of Group Actions on C*-Algebras. VIII, 371 pages. 1987.

Vol. 1275: C.A. Berenstein (Ed.), Complex Analysis I. Proceedings, 1985–86. XV, 331 pages. 1987.

Vol. 1276: C.A. Berenstein (Ed.), Complex Analysis II. Proceedings, 1985–86. IX, 320 pages. 1987.

Vol. 1277: C.A. Berenstein (Ed.), Complex Analysis III. Proceedings, 1985–86. X, 350 pages. 1987.

Vol. 1278: S.S. Koh (Ed.), Invariant Theory. Proceedings, 1985. V, 102 pages. 1987.

Vol. 1279: D. Ieşan, Saint-Venant's Problem. VIII, 162 Seiten. 1987.

Vol. 1280: E. Neher, Jordan Triple Systems by the Grid Approach. XII, 193 pages. 1987.

Vol. 1281: O.H. Kegel, F. Menegazzo, G. Zacher (Eds.), Group Theory. Proceedings, 1986. VII, 179 pages. 1987.

Vol. 1282: D.E. Handelman, Positive Polynomials, Convex Integral Polytopes, and a Random Walk Problem. XI, 136 pages. 1987.

Vol. 1283: S. Mardešić, J. Segal (Eds.), Geometric Topology and Shape Theory. Proceedings, 1986. V, 261 pages. 1987.

Vol. 1284: B.H. Matzat, Konstruktive Galoistheorie. X, 286 pages. 1987.

Vol. 1285: I.W. Knowles, Y. Saitō (Eds.), Differential Equations and Mathematical Physics. Proceedings, 1986. XVI, 499 pages. 1987.

Vol. 1286: H.R. Miller, D.C. Ravenel (Eds.), Algebraic Topology. Proceedings, 1986. VII, 341 pages. 1987.

Vol. 1287: E.B. Saff (Ed.), Approximation Theory, Tampa. Proceedings, 1985–1986. V, 228 pages. 1987.

Vol. 1288: Yu. L. Rodin, Generalized Analytic Functions on Riemann Surfaces. V, 128 pages, 1987.

Vol. 1289: Yu. I. Manin (Ed.), K-Theory, Arithmetic and Geometry. Seminar, 1984–1986. V, 399 pages. 1987.